VICKERS-MAXIM MACHINE GUNS

1886 to 1968 (all models)

COVER IMAGE: Gun, Machine, .303in, **Vickers Mk 1.** *(Author)*

Dedication

To all of those who served alongside Hiram Maxim's invention, in particular the men of the Machine Gun Corps and to Sgt J.W. Brooks DCM, MM Rifle Brigade and MGC, Gunner Leslie Gutsell, MMGS, and Lt Col Julius Caesar who were three of the first.

© Martin Pegler 2019

All rights reserved. No part of this publication may be reproduced or stored in a retrieval system or transmitted, in any form or by any means, electronic, mechanical, photocopying, recording or otherwise, without prior permission in writing from Haynes Publishing.

First published in November 2019

A catalogue record for this book is available from the British Library.

ISBN 978 1 78521 563 6

Library of Congress control no. 2019942929

Published by Haynes Publishing,
Sparkford, Yeovil, Somerset BA22 7JJ, UK.
Tel: 01963 440635
Int. tel: +44 1963 440635
Website: www.haynes.com

Haynes North America Inc.,
859 Lawrence Drive, Newbury Park,
California 91320, USA.

Printed in Malaysia.

Senior Commissioning Editor: Jonathan Falconer
Copy editor: Michelle Tilling
Proof reader: Penny Housden
Indexer: Peter Nicholson
Page design: James Robertson

VICKERS-MAXIM MACHINE GUNS

1886 to 1968 (all models)

Enthusiasts' Manual

An insight into the development, manufacture and operation of the Vickers-Maxim medium machine guns

Martin Pegler

Contents

6	Acknowledgements
7	Introduction
10	The machine gun concept
22	Hiram Maxim, trials and tribulations
34	The Vickers connection
46	First World War manufacture
68	Second World War manufacture and improvements
	Ammunition — 73
78	The Vickers in service
98	Vickers variants
	The US Model 1915 Vickers — 104
	The ANZAC Vickers — 107
	Canadian Vickers — 112
	The Russian Vickers-Maxims — 116
118	Maintaining and shooting a Vickers
146	A user's view
162	In retrospect
169	Appendix
	Vickers gun specifications — 169
170	Index

OPPOSITE Hiram Maxim in a gently lampooning cartoon by 'Spy' (Leslie Ward) from *Vanity Fair*, dated 15 December 1904. Years of testing machine guns had left Maxim very hard of hearing. *(Author)*

RIGHT Britain armed a large number of troops who had escaped from occupied Europe. These two gunners are Norwegians, setting up their Vickers gun prior to a live-firing exercise somewhere in Scotland, May 1942. The gun appears to be a refurbished First World War model. *(Imperial War Museum-IWM-H20090)*

Acknowledgements

No factual work of this type is solely the result of the author's endeavours, so I would like to thank the following institutions and individuals who have provided me with help:

The Royal Armouries and National Firearms Collection (NFC) in Leeds, who hold the finest collection of Vickers and Maxim guns to be found anywhere in the world. For anyone wishing to understand the development of these guns, I can do no better than suggest you visit the museum. If you wish to see a Vickers working, I recommend that you view the film made by them at: https://www.youtube.com/watch?v=84RCfcH_xuU.

I am also grateful to: Jill Dye, Information Officer at the Institution of Mechanical Engineers, London; The Vickers Archive, held by Cambridge University; The RAC Tank Museum, Bovington; Springfield National Armory, Springfield, Massachusetts; both the Rock Island Auction Company (website: www.rockislandauction.com) and International Military Antiques (website: ima-usa.com), kindly supplied excellent images of some of the rare Vickers guns that have passed through their hands.

Individuals who get a special mention include: Ian Skennerton for permission to use the parts lists and drawings from his Vickers book; Eugenie Brooks; Lawrence Brown; Simon Dunstan and George Yannaghas for supplying photographs; and Robert G. Segel for pictures of his Colt-Vickers gun; Ian Durrant for many photos and letting me prise 'Jean' from him; Peter Smithurst, Curator Emeritus at the Royal Armouries for all of the Maxim material; the late Herb Woodend, curator of the MoD Pattern Room (now part of the NFC collection) who gave me so much help and information in my early days of Vickers ownership; Dr Bob Maze, for all his help and some considerable ammunition expenditure; and Dolf Goldsmith, whose seminal work on the Vickers and Maxim machine gun stands as a testimony to the energy and brilliance of Hiram Maxim.

I would also like to give particular thanks to two people. Nick Felton for his superb photography of 'Jean', the Vickers gun used in the colour plates, and Richard Fisher, owner of the most comprehensive private Machine Gun Corps and Vickers collection in the world, who was able to find all the errors that had escaped me, as well as helpfully suggesting things I had not thought of and for photographing rare items from his extraordinary collection. His online 'Vickers MG Collection and Research Association' is a treasure trove of Vickers information and can be accessed at: www.vickersmg.org.uk. Thank you both.

Finally, a heartfelt 'thank you' to my long-suffering wife, Katie, who once again accepted with infinite patience and good grace the countless hours I spent torturing a keyboard, and who used her eagle eyes to perform editing surgery on what I thought was a perfect text. (It wasn't.) Despite her hard work, and that of the editors at Haynes Publishing, any errors or omissions that remain are entirely my responsibility. I hope the end result is a worthy tribute to the man and his machine.

A note on sources:

I was very fortunate to interview many Machine Gun Corps (MGC) veterans in the late 1970s and early 1980s, and I am indebted to them for sharing their experiences. They were a marvellous generation, now sadly all gone. I am also grateful to my late 'Uncle' Bill Cooke, H.L.I., for talking candidly about his war and for the Vickers lock, which I treasure. Several other sources have been used, many of which are anonymous, including personal diaries and regimental histories and I wish that there was space to include all of the extraordinary stories that I found.

Martin Pegler,
Val d'Issoire, France

Introduction

The history of the development of almost everything man-made is one littered with false starts, lost opportunities and dead-ends, frequently due to the inability of one person to make vital connections with the work of another. That ideas eventually came to fruition was often as a result of technology becoming available that had not been extant a few years previously, or because of a sudden mental illumination that guides an individual down a path hitherto never before taken. Sometimes, of course, it is no more than sheer, blind luck.

Generally speaking, making things work is a long process of adding together the sum total of work done by others and creating order from a random series of experiments that had, until that point, proven inconclusive. Rarely is a major invention the result of the work of a single individual working in seclusion; almost everything that comes to fruition is due to the hard work of others. As an example, the practical internal combustion engine came into being because of the work of over a hundred different inventors and engineers, across a time-span of 150 years. Inevitably, though, it falls to one person to piece together the complex jigsaw of disparate ideas to finally assemble the complete picture, but because history is neither fair nor accurate, they become labelled 'the inventor'. In reality, electricity was not invented by Michael Faraday, the aeroplane was not the result of a brief flight by the Wright brothers and Hiram Maxim did not invent the machine gun.

From the inception of firearms, it was the dream of soldiers, gunmakers and inventors to be able to create a gun that could fire more than one shot without having to be reloaded. With the primitive, single-chambered, black-powder-charged guns in use from the 14th to the 18th centuries, this was inevitably destined to be an unattainable concept. Slow-match or flint ignition, the vagaries of gunpowder and the need for a gun to be regularly cleaned of the sooty, choking fouling created by the discharge meant that even if a mechanism could be designed that would function efficiently, it was unlikely to have been of much practical use.

The first tiny breakthrough occurred in the early 18th century, when an English inventor named Puckle devised a large tripod-mounted repeating flintlock gun with a rotating magazine that could discharge up to nine shots in succession using a crank handle. Over the next century the idea was refined and miniaturised and was eventually to lead to the successful design of not the machine gun but the revolver, with which we are all familiar today. Nevertheless, the concept created sufficient interest for it to be followed through and over the subsequent decades a number of attempts were made to produce a more practical automatic gun. For the most part these were doomed to failure due to the inevitable requirement for black powder to be loaded into individual chambers and ignited by a fickle flint mechanism. The technical breakthroughs that

BELOW The crossed Vickers guns cap badge of the Machine Gun Corps. *(Author)*

were needed came, as often happens, hard on the heels of one another.

The first came in 1807 with the practical application of fulminate as an ignition, being invented by a Scottish minister, Alexander John Forsyth. It was not perfected until the early 1820s, when it appeared in its now familiar guise of a small copper cap. It was waterproof, powerful and rendered redundant the priming charge required to ignite the main charge. A firearm could now be loaded, capped and kept ready to fire in an instant, whereas gunpowder charges in flintlocks had to be removed on a daily basis if not fired, due to its propensity for soaking up moisture like blotting paper.

Once a reliable form of ignition had appeared, the next major advance was made in 1854 when Louis-Nicolas Flobert created a self-contained copper cartridge with a .22-calibre lead ball seated on top. Although tiny and only just powerful enough to pierce paper at 6ft, it was quickly adopted by Horace Smith and Daniel B. Wesson in 1857 for their new range of revolvers. So successful was it that their .22 cartridge is still in production today.

These incremental technological steps may appear to be irrelevant to the development of the machine gun, but they laid the foundation for the route that eventually led to the creation of types of firearms that could chamber the new metallic cartridges. More crucially, this meant that they could finally be applied to automatic weapons, providing a reliable ammunition supply contained in separate magazines without the unpredictability of

BELOW The perfectly attired King's Royal Rifle Corps, MG Section, Chitral, India, 1895. The gun is a M1893 Service Maxim in .303 calibre and is mounted on the short-lived heavy Mk I tripod. *(Author)*

requiring individually loaded chambers. Therefore, it is hardly surprising that in the mid-19th century several manufacturers simultaneously began work on prototype machine guns. In France in 1859 Joseph Montigny produced a 50-barrelled, cannon-sized gun called *le mitrailleuse*. It appeared during the Franco-Prussian War of 1870, but was employed as artillery, albeit possessing insufficient range to be effective. In 1861 the American-designed Ager gun was demonstrated to President Lincoln but suffered from a number of deficiencies.

It is thus somewhat ironic that the first workable machine gun should have been invented by a doctor of medicine, R.J. Gatling, in 1862. Being hand cranked enabled the speed of fire to be carefully regulated to prevent overheating and the Gatling, in various guises, went on to become widely used both by the US Army and Navy and also by Great Britain's land and sea forces. Indeed, electric-powered variants are still much used as aircraft and helicopter armaments today. It too had inherent difficulties, the most crucial being that the design failed to address the single most fundamental problem in creating a truly automatic firearm: how to automatically re-cock the action after every shot and not have to rely on manual operation. The firing of a cartridge creates a massive amount of wasted energy in the form of heat that is seen as muzzle flash. This can be as much as 67% of the total energy produced by the powder charge, so finding a method of harnessing it efficiently was a problem that many engineers and firearms designers wrestled with, usually unsuccessfully. But there is an exception to every rule and, in this case, it was a quiet mechanical genius called Hiram Maxim.

BELOW Maxim was a great publicist, and provided two Perfect Models for use in the play *A Life of Pleasure* at the Princess Theatre, London, in 1897. They filled the theatre with powder fumes, deafened the audience and were described as 'a resounding success'. *(Vickers Archive)*

Chapter One

The machine gun concept

A look at where the idea of rapid-fire came from and why it was regarded as so important, and a fascinating glimpse at a few examples of the most significant inventions that eventually led to the production of the first repeating-fire guns – with perhaps just a few false starts along the way.

OPPOSITE The only extant photograph of Maxim's stand at the International Inventors' Exhibition in 1888 where his guns were awarded a gold medal. Both guns still survive and are displayed at the Royal Armouries Museum, Leeds. *(Author)*

The function of early guns was limited by both the available technology (frankly, there wasn't much) and the propellant – gunpowder or black powder as it is more commonly known today. A gun of any type, be it cannon or small arm, comprises three main elements: a tube or barrel, a method of ignition and a projectile. Small medieval guns, known as handgonnes, were no more than the sum of those parts. There was a crudely cast bronze or pattern-welded smooth-bored iron barrel, usually of large bore, perhaps a 1in (25mm) or larger, with a breech-hole for ignition. They were often of considerable weight, between 10 and 20lb and with no sights or stock they could not be aimed, so were accurate only out to a limited range, perhaps 25yd or so. But in a world where fighting was done by means of sharp weapons – swords, lances and arrows – their appearance on the battlefield, with the consequential smoke, flame and incredible noise, would have been terrifying to both men and horses alike.

As with all inventions, the passage of time saw incremental improvements in the form and function of the gun, and there naturally were a few blind alleys along the way, such as the complex, elegant and expensive wheel-lock of the early 16th century. It sometimes worked when it was required, but it was a brave man who trusted his life to it. Better in all respects for the future concept of the machine gun was the snaphaunce, which appeared in the mid-16th century and had a cock, steel pan containing the priming powder and a frizzen, a steel plate against which the jaws of the cock, holding a piece of flint, sparked. It was no more than a mechanical tinder-lighter, with the addition of a relatively efficient trigger mechanism. By the first decade of the 17th century, this had been refined into the familiar flintlock mechanism, which was to remain the primary arm for both military and sporting firearms for the next 200 years.

The problems with the flintlock, and indeed all of the types of firearm mentioned so far, were many. The gun had to be loaded, primed and hand-cocked for each shot. They were unreliable in wet weather, to the extent that battles were seldom fought in strong winds or heavy rain, as the likelihood was that around 80% of the muskets would fail to ignite. In addition, gunpowder was possibly the worst form of combustible material that could be conceived, it being hygroscopic. It acted like a sponge in damp conditions, as well as being highly corrosive to all metal surfaces. If that were not enough, it generated 300 times its own volume in dense white smoke when fired, totally obscuring the target and immediately giving away the presence of any shooter who was concealed. The victor of the battlefield in the years of linear warfare was often the side that had the wind behind it. These limitations did not, of course, deter the many inventors who tried to improve the functioning of these primitive firearms, but they could only work with what was available.

Early types of repeating guns did begin to appear: an English snaphaunce revolver exists in the Royal Armouries and National Firearms Collection that dates to around 1670 and flintlock revolving pistols and muskets made by Elisha Collier of Boston can be found in several museum collections, all dating from the beginning of the 19th century. Other, more esoteric, solutions involved multiple-barrelled pistols (quite literally, the first revolvers) whose barrel cluster moved round with each successive shot – the same basic concept as the Gatling gun.

Alas, bigger guns such as muskets with rotating barrels were simply too heavy to be practical but one attempt to use the multiple barrel concept was famously produced by the eccentric English inventor James Puckle who

BELOW **Puckle gun, showing the large crank handle, flintlock priming mechanism on top and the bizarre square chambers. The lettering on the breech says: 'A Defence'.** *(Courtesy Trustees of Royal Armouries Museum)*

designed his Defence Gun in 1718. These curious weapons were of iron or bronze, 3ft long, mounted on a heavy tripod and weighed 78lb. They had a rotating six-chambered cylinder that, depending on the preference of the firer, could shoot round balls at European enemies, or square shot at Turks, whom Puckle detested for no obvious reason. The cylinder was turned by a crank handle, and each chamber had to be manually realigned with the breech before it was discharged. Having only a single flintlock mechanism it also had to be reprimed for each shot, but despite this, it attained a rate of fire of 63 rounds per minute (r/pm) when demonstrated to the British Board of Ordnance in 1722, using preloaded cylinders. The Board remained resolutely unimpressed, however. What is particularly interesting from the point of view of automatic weapons development is that when entered on a ship's manifest for shipment to British forces in St Lucia, two Puckle guns were listed as '2 Machine Guns of Puckles', the first time the term was ever recorded. But for all its amusing elements, the guns were arguably the first true attempt to generate a level of firepower that was considered impossible for an enemy to confront. It is interesting that Puckle himself determined that the gun's primary function was '. . . that for defence'. The idea of offensive use for a machine of this nature was still an utterly alien one.

For the next century, little work was undertaken on the development of such weapons. The demands that needed to be met were more prosaic. Sporting and military guns were wanted that were faster to load, more accurate and more reliable. Clearly there were limits to the speed with which a muzzle-loader could be recharged; four shots a minute for a smooth-bore or two for a rifled gun were about average, with the added complications of barrel fouling making loading a progressively slower procedure and the requirement to change a flint once its edge became dulled. Yet the solution to muzzle loading had been in existence since the 16th century, in the form of breech-loading and several guns had been produced that utilised the idea.

In 1537 a breech-loading matchlock carbine was made for King Henry VIII and a more sophisticated snaphaunce musket was manufactured for Philip V of Spain in 1715. Both of these guns came with steel breech-blocks that could be preloaded in the manner of a cartridge. In 1772 a British Light Infantry unit under Major Patrick Ferguson was equipped with a different form of flintlock breech-loader that utilised a screw-breech system made by Durs Egg of London, the first such rifles ever to be adopted for military service. However, these were rare and horribly expensive. Besides, the biggest hindrance still lay with the method of ignition, and finding a solution to that was something that kept a myriad gunmakers awake at nights.

It would be all too easy to consign these inventions to the lost-luggage office of history, but they contained many elements that would,

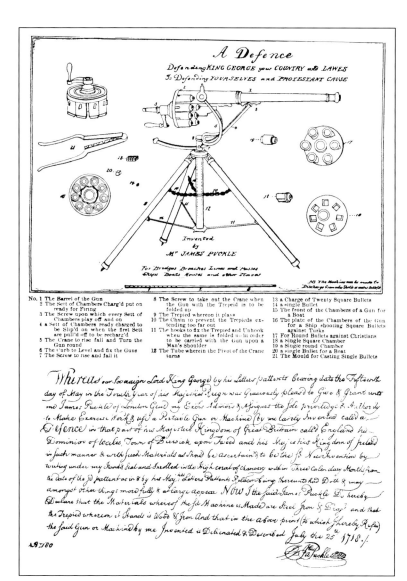

ABOVE Puckle's patent of 15 May 1718. It shows the gun and tripod with both round and square-chambered cylinders. The crank handle for chamber rotation is fixed to the rear of the cylinders. A multiple bullet mould is also depicted.
(Royal Armouries)

ABOVE Revolving flintlock by T. Annely, c.1710. A compact design, only 12in in length, with eight chambers, each of which (theoretically) lined up with the barrel by the action of cocking. In practice, when this failed to happen the result was a potentially lethal explosion.
(Royal Armouries)

in the next century, become familiar to those who designed automatic guns. Rotating barrels, crank operation and separate loaded cartridges were all ideas that would surface again. Although they initially had little effect on overall weapons design and construction, they did prove, albeit in a basic way, that the concept of rapid fire was at least viable.

As is so often the way with successful inventions, the early development work was done not by one individual, but a succession of people over a relatively long period of time. So it was with the development of fulminate of mercury. It was nothing new, a chemical composition that was so sensitive it was capable of destroying a firearm if used in large quantities. It had probably first been discovered in 1776 by Claude-Louis Berthollet who realised that chlorate could be exploded with a sharp blow, although finding a practical application for it was something else entirely.

It fell to a dour Scottish clergyman and duck-hunter, the Reverend Alexander John Forsyth, to appreciate that while it was utterly inappropriate as a propellant, fulminate was the perfect composition to be a priming compound. In 1807 he developed the Scent-Lock, so named from the shape of a container that deposited a tiny amount of fulminate primer on to an anvil, which was then struck by the hammer, igniting the main charge. Aside from the occasional unfortunate explosion it worked tolerably well, but it then took an ex-patriate Englishman living in America, Joshua Shaw, to refine it further. He manufactured a small brass cap, resembling a top-hat, and filled the inside with a small quantity of fulminate, thus creating the now-familiar percussion cap. Placed on a hollow nipple that was screwed into the breech of a gun and ignited by the hammer, it totally revolutionised firearms design. From its introduction in around 1820 it took no more than 20 years to supplant the old flintlock.

Soon a new type of gun appeared, the percussion (or capping) muzzle-loader, and it was adopted wholesale by the armies of the world, who quickly converted existing stocks of now-redundant flintlocks. It was weatherproof, kept the breech sealed from moisture, could be left in place indefinitely and worked instantaneously without any time-lag or creating a tell-tale plume of ignition smoke.

There were other advances, too, for the Industrial Revolution had begun to change the science of metallurgy and improve methods of manufacture. Gun barrels, for example, became cheaper, stronger and more accurate, as rifling, once a highly skilled and laborious process, became mechanised. Smooth-bore muzzle-loaders were slow, inaccurate (80yd was about the maximum range to be able to guarantee a body hit with a military musket) and windage (the escape of propellant gas past the ill-fitting round lead ball) meant that projectiles lost velocity with alarming rapidity.

But aside from the few rare exceptions already mentioned, cartridges hitherto had been large steel chambers or fragile combustible paper, but in 1808 a Swiss artilleryman and engineer, Jean Samuel Pauly, invented a simple, small self-contained brass cartridge that was further perfected by Casimir Lefaucheux in 1836. This had happily coincided with increasing interest in breech-loading, the long story of which is not one to dwell upon here. Suffice to say that after several false starts, and one or two promising ones, the practical breech-loader appeared as a result of the usual reason – war. The American Civil War produced such a plethora of designs that entire books have been devoted to them. When it began in 1861, flintlocks were still commonplace, but by the time it ended in 1865, repeating cartridge breech-loaders such as the Spencer, Henry and Sharps were widespread.

By the late 19th century, gunmakers and designers had not just one, but literally dozens of designs to choose from and the demand for metallic ammunition had soared. Then, an advance in metallurgy occurred that was about to turn the firearms world on its head. One of the greatest frustrations for gunmakers attempting to produce rapid-fire weapons was the need for separate charges of powder and a primer. Previous attempts to produce self-contained cartridges were hampered by the lack of a material that could form the case for a cartridge. Brass would be the ideal material, being plentiful, malleable and capable of withstanding high temperatures, except there was no available method of forming or drawing it into the required shape. To an extent this was possible with copper, though, and in 1845 Flobert's tiny 6mm lead bullet, contained in a copper primer, was in many respects a quantum leap forwards. Its performance was totally underwhelming, however, so in 1857 Horace Smith and Daniel Wesson seized the idea and refined it to produce the .22-calibre Short for use in their new bored-through cylinder revolver, the Smith and Wesson No 1. By feeding a thick brass cup through a series of drawing presses, and annealing it to soften the metal, it was possible to stretch it into a thin, strong cylinder, with a solid base that was filled with fulminate primer. This resulted in a totally waterproof, unbreakable cartridge with a self-lubricating bullet that, given reasonable storage conditions, had an almost indefinite shelf-life. It was the breakthrough that the firearms industry had been waiting for, and it is only a slight exaggeration to say it changed the evolution of weapons design almost overnight. Very few people realised at the time just what a significant advance had been made as a result and what the repercussions would soon be for the firearms industry and the development of the machine gun in particular.

So, where did that leave the small number of gunmakers attempting to create a gun that could reliably operate on a semi-automatic, or automatic basis? Most, it must be said, were still at a complete loss. Although cartridges now existed, as did efficient breech-loading, the physical constraints of uniting the two into something that was workable (and did not look and perform like a large piece of artillery) seemed beyond almost everybody. This, of course, did not stop inventors and engineers from trying.

Arguably the first workable automatic gun was produced by Sir James Lillie in 1857 whose 'Battery Gun' comprised of two rows of six barrels, each with a revolving cylinder that contained 20 loaded chambers. That familiar item, a crank handle, worked a complex series of mechanical arms that acted as hammers. Inevitably, each chamber had to be manually aligned before firing and once empty, the 'cartridges' had to be removed for reloading. It was neither practical nor effective as well as being so heavy as to be almost immobile. Despite being demonstrated at Woolwich Arsenal in 1857, the project was stillborn and the gun still resides in the Royal Artillery Museum at the Rotunda in Woolwich.

The most prolific firearms inventors were to come from America, where the machine gun concept seemed to have struck a chord. The British Army was steeped in tradition, reluctant to adopt new technology and had generally fought and won wars against colonial enemies

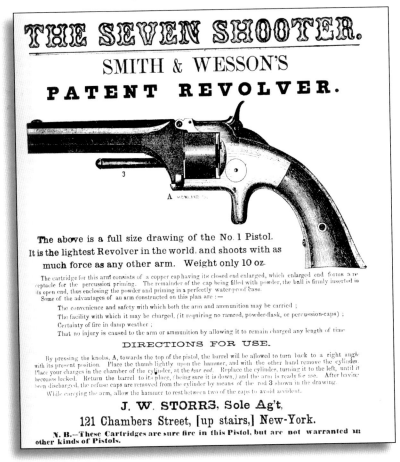

BELOW An advertising flyer from c.1859 for Smith and Wesson's revolutionary new 7-shot No 1 .22 calibre revolver. Pressing the latch 'A' allowed the barrel to hinge upwards enabling the cylinder to be withdrawn, and the fired cartridges were removed by using the fixed rammer '3'. It was to revolutionise revolver design. *(Author)*

who were poorly armed and indifferently led. New and potentially expensive technology was of little interest as there was almost no military requirement for it.

But the experience of the Union and Confederacy on the American battlefields had been very different to that of the British, and technological advancement in small-arms design had been a major factor in determining how the US Civil War had been fought and eventually won. As a result, the United States Army was more open-minded about new ideas, with financial backers more likely to invest in the new technology, and this was to provide greater encouragement to inventors and engineers alike. In the decade prior to the Civil War, several American designers had come up with ideas that, if not totally viable in themselves, helped to pave the way for later development. The most notable of them were Dr Josephus Requa, Charles Barnes, Ezra Ripley, Wilson Agar and Richard Gatling.

Requa, a dentist by profession, had worked in his youth for gunmaker William Billinghurst and retained his fascination for firearms for the rest of his life. In 1858 he had read an article in *The Engineer* magazine about the work of Lillie, and believed that he could improve upon the design, so in 1860 he contacted Billinghurst with an idea for a bank of 25 rifled .58-calibre barrels mounted in line on a single wheeled carriage. They produced a working model by July 1861 that had a loaded strip magazine containing 25 paper cartridges in steel tubes, which were fired simultaneously by a trigger. More sophisticated than the Lillie, the barrels could be aligned depending on the range, allowing them to provide diverging or converging fire. It was not, strictly speaking, a machine gun, but a battery gun. Unlike the Lillie, however, it was easily portable, served by a crew of three and was capable of firing at 175r/pm. During US Board of Ordnance testing it proved accurate and quite devastating out to 600yd. Despite strong objections from the ultra-conservative Chief of Ordnance, General J.W. Ripley (overridden by President Lincoln), 50 were ordered. Although never officially adopted into service, the gun did see considerable combat in the latter stages of the war, and was considered to be a reliable weapon. But it proved to be the last gasp for guns of this type, for metallic cartridges had become commonplace by 1865, rendering fixed multiple-barrelled guns redundant.

There were still more complex designs on the drawing board, however. C.E. Barnes was a gunmaker living in Lowell, Massachusetts, who was fascinated by the concept of automated fire. He created what he termed 'an improved automatic cannon' which received US Patent No 15,315 on 8 July 1856. In many respects, it contained all of the disparate elements needed to produce a workable machine gun. Although it too had to be hand-capped with a percussion nipple before firing, it had a loading tray that contained hard-bodied linen cartridges, and the single barrel was loaded by means of a strong but simple toggle-joint, which functioned in much the same manner as a human arm or leg joint. Barnes harnessed some of the recoil pressure generated by the propellant gas to actuate the hammer, moving it into the cocked position after each shot. Both of these ideas were later to resurface in the designs of Hiram Maxim. It was hand cranked and not overly efficient as the rate of fire was slowed by the continual need to cap the chambers prior to every shot, but it was certainly a step in the right direction.

At the same time another gunmaker and inventor named Ezra Ripley (no relationship to General Ripley) of Troy, New York, was working on a more advanced design that did not rely on a single fixed barrel, but a series of nine rotating ones. This had the benefit of permitting each one to cool after firing and although this also required each barrel to be individually loaded, Ripley followed tradition and had preloaded and capped steel 'cartridges'. Although patented in 1861, it seems unlikely that any working example was ever made.

The next incremental advance in the chain of developments was a gun designed by Wilson Agar (sometimes spelled Ager). He is a shadowy character; other than a listing in the British patent office as 'an inventor', little is known about him. In England in 1860 he had patented the design for his Union Repeating Gun, filing another patent the following year in the US Patent Office. Like Barnes's gun, this too was a multi-chamber single-barrelled gun operated by the inevitable hand crank. However, its novelty was in using a vertically mounted hopper to hold preloaded steel tubes each holding a paper cartridge of standard .58 calibre. These hoppers looked very much like commercial coffee-grinders, hence the nickname later applied to Agar's guns. The handle was pulled back and pushed forwards which operated a wedge that cocked and fired the cartridge, the empty case dropping into a pan underneath. As long as sufficient ammunition could be provided, it fired at a rate of 100r/pm. This meant that in the space of a minute 7,500 grains of powder was combusted in a single barrel. An observer, Major G.V. Fosbery VC (himself a talented gun designer), pointed out that that was its undoing: 'Seven pounds of lead would pass through this single barrel in that time. The effect [. . .] proved to be that the barrel grew first red hot, then nearly white hot, large drops of fused metal poured from the muzzle, and firing had to be discontinued.' A steel shield was also thoughtfully supplied to protect the gunners, a practical idea not adopted again until the First World War. The gun so impressed President Lincoln (most new inventions appeared to) that in 1861 he ordered 64 of them to be supplied to Union forces, though they saw little action throughout the war and never really fulfilled their potential.

All of these contemporary designs comprised a mix of advanced thinking limited by the vision of their inventors and the technology available. Still, elements of the true machine gun were appearing: recoil assistance, toggle locking, magazine feeding and cooling systems now existed, but not in any single design. Of course, many other firearms designers had also been hard at work post-1865, each trying to perfect the machine gun concept, with varying degrees of success. Most sank without trace, but three others deserve further mention as each advanced the concept and were adopted into military service by several countries.

William Gardner of Toledo, Ohio, had fought through the American Civil War in the Union Army and, having witnessed some of the multi-shot guns in action, was convinced that a practical automatic design was feasible. To avoid overheating, he opted for a .45-calibre two-barrelled gun using a complex but efficient hand-cranking system, with the added benefit of a pair of wedges he called 'shell starters'.

ABOVE Gardner and Nordenfelt guns were widely adopted throughout Europe. This is a Dutch two-barrelled M1879 .45-calibre Gardner on a fortification mount at Fort aan Dem Ham in the Netherlands. *(Author)*

BELOW A four-barrelled naval Nordenfelt, in 1in calibre on a pintle mount. The large hopper held the gravity-fed cartridges and the gun was fired by working a lever to and fro. *(Courtesy Trustees of the Royal Armouries Museum)*

These positively assisted the ejection of the fired case, as one problem with early metallic ammunition had been a tendency for the base of the soft cartridge to be ripped off by the extractor, leaving the fired case stuck firmly in the breech. Although rejected by the US Army, Gardner took his design to England, where an example, tested at Portsmouth Naval Dockyard, fired an incredible 812r/pm. The Royal Navy adopted it, and it first saw combat in the Sudan in 1884, where its firepower proved devastating.

After this, designs came thick and fast in England. Three Swedish engineers, Helge Palmcrantz, Johan Théodor Winborg and Eric Unge, had developed and patented a rifle-calibre light gun in 1873, named the Nordenfelt after its financial backer, Thorsten Nordenfelt. It was a multi-barrelled design, using between five and ten barrels, and during British naval tests in 1882 it managed a very impressive 1,000r/pm. Bearing in mind the action was worked by the sweating gunner moving a vertical lever to and fro, this was even more remarkable. It was subsequently adopted by both the Royal Navy and the Army in a variety of calibres from .303in to 37mm. Curiously, his most advanced design, a lightweight single-barrelled gun weighing just 13lb and crewed by two men, never went into production. Nordenfelts were subsequently to become Maxim's closest rival, and they do not disappear from these pages quite yet.

This early development work, plus the breakthrough in metallic ammunition, enabled a North Carolinian named Richard Jordan Gatling to continue where others had left off. He was neither an engineer nor a gunmaker, but a qualified medical doctor, albeit one with an inherent mechanical interest. He understood the limitations imposed by the single barrels as well as the problems generated by an inadequate ammunition supply and in 1862 he patented a multi-barrelled gun with six .58-calibre rifled barrels that rotated around a central hub.

He used a vertical hopper, into which were loaded preloaded and capped steel cartridges, in much the same manner as the Agar. The mechanical loading and firing system was ingenious, employing a helical cam that provided a reciprocating motion to the locks. Behind each barrel was a cartridge carrier and behind these were the lock housings, the barrels being revolved by gears operated by a revolving hand-crank handle and all of these elements were secured to the central shaft so that there was no possibility of their rotating independently. This also ensured that the lock housing and cartridge carrier were in battery (correctly lined up with the breech) before firing. As the barrel continued to revolve, the bolt moved forwards and once locked into place it fired the round. There was a slight delay built in to deal with hang-fires, and as it continued to revolve, the bolt opened and ejected the fired case. This arrangement had added benefits: as the hot case was removed, the open chamber allowed cooling air to pass through the barrel, as well as simply bypassing any stoppages caused by unfired ammunition – something that plagued single-barrelled guns.

One major problem that Gatling faced was being unable to solve obturation, the leakage of gas through the unsealed gap between the breech face and cartridge. Luckily salvation was at hand in the shape of the new .58-calibre copper rimfire cartridges that had become commonplace by the end of the Civil War. In 1866 he converted his guns to shoot rimfire ammunition and the soft cartridge base acted as a gas seal, while additionally providing greater velocity and improved range.

Using metallic ammunition called for a radical redesign of the mechanism, however, as the cartridge base now had to be firmly supported by the lock upon firing. Problems with case ejection had to be resolved and the side effect of this increased efficiency was heat build-up in the barrels that Gatling attempted to solve by devising a water-cooling jacket, although this never appears to have been put into production. All of these problems were to be faced by Hiram Maxim some two decades later, and he would re-examine Gatling's work with a critical eye.

Successful tests by the US Army Ordnance at Washington Arsenal in January 1865 had resulted in the Gatling being adopted for service and manufacture was duly taken up by the Colt Patent Firearms Manufacturing Company. In 1871 Gatlings were standardised to accept the US Army's new .45-calibre metallic ammunition and Britain also adopted its first machine gun, a ten-barrelled Gatling in .45-Martini-Henry calibre, first using it in Peru in 1877.

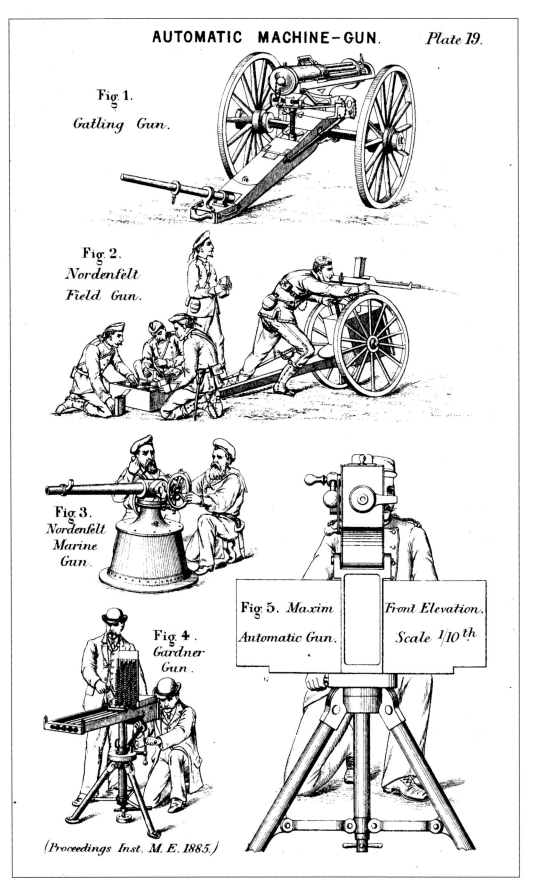

LEFT An illustration from *The Engineer* used to demonstrate Maxim's lecture in London in 1855, showing the relative compactness of this gun compared to his rivals. *(Peter Smithurst)*

RIGHT A model 1862 Gatling, modified to chamber .58-calibre rimfire cartridges. The ammunition hopper sits on top of the breech, with the firing crank handle on the left. It was made in Cincinnati by McWhinney, Rindge and Co., in 1863. *(US Armory National Park Service, Springfield Armory National Historic Site. Photo James Langore)*

Gatling had realised that the rate of fire of any machine gun was effectively limited by the amount of ammunition that could be loaded, so a drum feed devised by James Accles was adopted in 1883, which enabled the rate of fire to be increased to 300r/pm. But the fact that these guns were still hand cranked was a limiting factor, for a human arm can only work so fast. Early experiments were therefore made with an electric motor, which proved surprisingly effective, generating a rate of fire of around 1,200r/pm.

RIGHT Ammunition fired in Vickers-Maxim guns. From left: .450in Martini-Henry, brass foil case; .45 Gardner-Gatling, solid brass case; .45 Maxim, Mk VII .303in. *(Courtesy George Yannaghas)*

The remaining problem, though, was that most guns were to be found in places where electrical power was not commonplace, and heat build-up led to 'cook-offs' (where the chamber becomes so hot that cartridges self-ignite). Gatling is still feted by history for inventing the first machine gun, but in reality he had not solved any of the major problems that plagued designers: the waste of the propellant gas, hand cranking the firing mechanism, automatically reloading the gun and miniaturising the entire ensemble, for the Gatling gun weighed 170lb in addition to its wheeled carriage.

By now cartridge ammunition was commonplace, but gunmakers were still hampered by the limitations imposed on them by the black powder they still contained. Because gunpowder burned relatively slowly, sizeable charges were needed along with large-calibre bullets, as the low velocities required a projectile that possessed a significant mass to impart the 'knock-down' impact required to ensure it was lethal, hence calibres of .45, .50 or 1in. A typical .45/70-calibre US military rifle cartridge had a lead bullet that weighed 400 grains and required around 70 grains of powder to propel it at 1,550 feet per second (fps). This size reduced the quantity of ammunition that could be carried, as well as placing limits on range and accuracy due to the rapidity with which the bullets lost velocity and were affected by crosswinds.

Early centrefire ammunition came in two types, British Boxer or American Berdan. Both had a centrally placed percussion cap, which was forced on to an anvil when struck by the firing pin, but early Boxer cases for military ammunition were of wrapped brass foil with a steel base crimped in place, while Berdans were generally of all-copper construction. Both types were easily deformed, and the gun's extractor frequently ripped the base from the cartridge body (particularly the Boxer type) as it removed it. Though this appeared to be an apparently unsolvable problem, salvation was at hand in France.

The solution was found in 1884 by a French chemist named Paul Marie Vieille who was employed by the Laboratoire Central des Poudres et Salpêtres in Paris. He had been working on improving earlier experimental work with propellants undertaken in England and Austria, using soluble nitrocellulose that had been mixed with ether and paraffin. His invention was a white powder, called 'Poudre B' (Poudre Blanc), and it was to revolutionise small-arms performance and ultimately lead to the development of weapons systems that hitherto had been impossible. It was fast burning, three times more powerful than black powder and produced virtually no smoke on discharge, nor was it so horribly corrosive. In its original form it was a white granular powder, but was soon being manufactured in thin dark-grey sheets that were cut into tiny flakes. It was also waterproof and could not detonate unless compressed, making it very safe to handle. If accidentally ignited, it simply burned with a bright flame, but would not explode.

The French Army immediately appreciated the significance of this breakthrough, for it made large-calibre bullets redundant; rifles could now fire smaller calibres, at much higher velocities, with greater accuracy, and soldiers could carry more cartridges pound for pound than ever before. They lost no time in adopting a new bolt-action magazine rifle, the Fusil Modèle 1886, or Lebel, as it was generally known. It chambered a bottlenecked 8mm cartridge, the original bullet being a flat-nosed solid bronze pattern, replaced in 1898 by a uniquely streamlined pointed or 'spitzer' variant with a boat-tailed rear that improved its stability and range. Using 45 grains of nitrocellulose powder, this provided the 198g bullet with a velocity of 2,400fps and enabled accurate shooting to ranges in excess of 1,000yd – almost unparalleled among military rifles.

The importance of Vieille's powder, and the subsequent developments that were to result from it, cannot be underestimated, for it began a new chapter in the history of modern ballistics and it remains to this day the primary form of propellant for small arms. In an indecently short time, most of the world's nations had rearmed their soldiers with rifles converted or manufactured to fire the new ammunition. Gatling produced a variant of his gun in 1893 that chambered the new .30–40-calibre Krag cartridge, which was the first small-calibre high-velocity cartridge adopted by the United States military.

So what was the net result of this plethora of early development work with gun configurations, mechanisms and ammunition? The machine gun had clearly not yet sprung fully formed from the blueprints of any designer, engineer or company, but designs had coalesced with each other. Systems were borrowed, tested and discarded, but occasionally adopted and refined, so that by the early 1880s the basic foundation of what was needed to create a workable automatic weapon was in existence. But there were still significant problems to overcome. The first was that of hang-fires when a cartridge failed to ignite. If the gun barrels are being rotated at high speed, the chamber of the barrel could then be open when the cartridge explodes. The result of this can be catastrophic; the breech could be blown out, injuring the gunner, the magazine ignited and depending on the mechanical system, the rotating crank might drive a fresh round into a breech still containing the exploded one, locking the mechanism solid. The guns were also too heavy, had little ability to be traversed and usually required a two-wheeled carriage as well as an ammunition limber, several crew members and a team of horses. The situation needed someone with the ability to turn these inchoate ideas into a cohesive concept. More than that, they needed the skills and vision to turn mere ideas into working metal. That someone proved to be Hiram Stevens Maxim.

Chapter Two

Hiram Maxim, trials and tribulations

How did a self-taught engineer from rural Maine become the first man to overcome the myriad obstacles that had baffled professional gunmakers for almost two centuries? In the space of just a decade Maxim designed and built the world's first truly automated machine gun, and changed world history in the process.

OPPOSITE Hiram and his rarely photographed brother Hudson with a 'World Standard' carriage-mounted gun. Hiram is pointing to the lock assembly that he is holding. The long top-cover of the gun is open with the rear sight clearly visible above the shield, to the right of Maxim's shoulder. *(Vickers Plc)*

Maxim was born on 5 February 1840 in the rural village of Sangerville, Maine, the youngest of eight children. His father, a talented engineer, ran a grist mill but Hiram was a determined, restless young man, who did not take easily to formal schooling and had no intention of becoming a mill worker. He became an apprentice to a coachbuilder at the age of 14, and by 16 he had produced his first invention a 'self-powered mousetrap'. By all accounts it worked perfectly, capturing the mice alive, but in the years prior to the concept of mass-production it proved too costly to manufacture. At the age of 24 he went to work for his uncle Levi as an instrument maker and draughtsman and this provided him with additional practical skills to add to those he already possessed. He had both an enquiring and inventive mind, and a remarkable ability to understand mechanical concepts by visualising them in three dimensions, seeing things not as individual processes, but part of a cohesive system. Brought up in an era where every rural family had guns for hunting or self-defence, he initially showed little personal interest in firearms. He recalled in a lecture that he gave in 1854:

My father conceived the idea of making a machine gun. As metallic cartridges had no existence, he proposed to load short sections of steel tubes, each provided with a percussion cap, and to connect them together in the form of a chain. He proposed a single barrel, and to feed the cartridges into position by the action of a hand lever. The drawing back of the lever cocked the hammer and turned a star-shaped feed-wheel sufficiently to bring a loaded tube in front of the barrel. The forward action of the handle brought suitable mechanism to bear on the end of the tube, and by means of a toggle-joint, the tube was [. . .] pressed so firmly into the barrel that no gases could escape. He believed that with a gun of this kind he would be able to fire one hundred rounds in a minute from a single barrel.

This early outline from Maxim is notable in that it raised many of the problems that would beset later inventors. It is interesting, however, to note that he makes such an early mention of what would become a fundamental part of his own later design – the toggle-joint. As it happened, the gun never became a reality, for despite Hiram making a wooden pattern and engineering drawings of it, his uncle believed that it would cost about $100 to manufacture (in today's money, roughly £2,300) and pointed out that '. . . when finished would not be worth 500 cents'. In fact, his uncle was perfectly correct; considerable funding would be required to manufacture such complex prototypes and, as has been shown, this was only the beginning of a hugely expensive spiral of testing and development work. There was one interesting spin-off from this, though, for it increased Hiram's interest in firearms, and when he was offered the chance to shoot a regulation .58-calibre Springfield rifle, he commented: 'I found the musket gave me a very powerful kick, in fact, a great deal more than I expected.' This set his fertile mind thinking, for he realised

BELOW Hiram Maxim, aged 17.
(Maine Public Library)

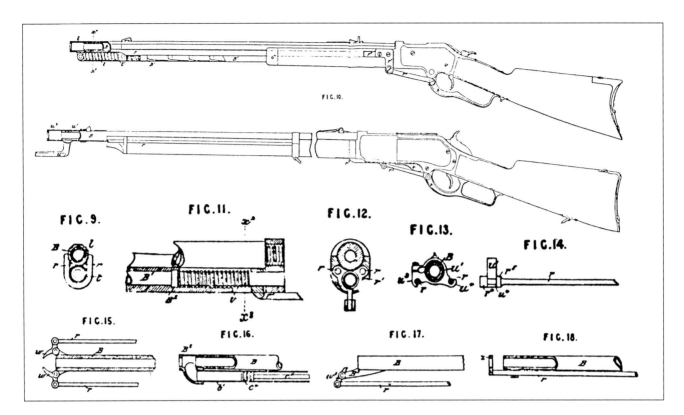

ABOVE Maxim's crucial Patent 606 of 1884, showing how he intended to trap waste gas to recock the mechanism. Although it was an auto-loader, not a machine gun, the concept was still a valid one.
(Peter Smithurst)

that the energy of the cartridge, once it had propelled the bullet from the barrel, was simply wasted as a gaseous flame – the muzzle flash – that vented into the atmosphere when any gun was fired. In his autobiography he noted:

> I related my experience to my father. I told him that I believed the energy in the kick of a military rifle would be amply sufficient to perform all the functions of loading and firing, so that if the cartridges were strung together in a belt, a machine gun might be made in which it might only be necessary to pull the trigger, when the recoil would feed the cartridges into position, close the breech, release the sear, extract the empty case, expel it from the arm and bring the next into position.

This idea was extraordinary, more so because he visualised the entire system as one holistic mechanical function: extract, chamber, fire, eject, extract, etc. It was even more surprising as he was still such a young man and had no formal gunsmith training or even much firearms experience. By this date, around 1870, the Gatling had become the premier machine gun in the US Army's arsenal, and Maxim was well aware of its shortcomings. But that mattered little at the time, for he had to earn a living by some means, and designing expensive automatic firearms was clearly not going to be one of them.

Maxim is, of course, known today solely for his work with machine guns, but this happened only because of his success with a different technology that could not be further from that of firearms. He was utterly fascinated by the science of electrical energy, and worked tirelessly to produce apparatus that had practical, profitable applications. There is no need to go into great detail about his work, suffice to say that his genius appeared to know no bounds. He designed an improved carburettor, a water heater, steam trap, automatic pumps, one of the first pressurised fire extinguishers and electrical hair curlers. Between 1866 and 1884, of the 83 patents he filed, no fewer than 43 were to do with electrical equipment. Neither should this be dismissed as dabbling in peripheral work within already well-established sciences, for these were serious inventions in their own right. He patented a current regulator and produced one of the first electric dynamos, but by far the most important invention relating to his future work was the light bulb, for the fact is that it was *not* Thomas Edison who perfected the electric lamp as we know it today, but Hiram Maxim. Maxim beat Edison by a year to perfecting the incandescent electric light when he solved the twin problem of how to remove air from light bulbs, which caused the filaments to burn out, while simultaneously perfecting the manufacture of long-burning carbon filaments, which he wisely patented. The problem for Maxim was that Edison was backed by the immensely wealthy and powerful United States Electrical Lighting Company, who had invested very heavily in his work over many years and understandably wanted to have a controlling interest in the lucrative supply of electric light and manufacture of light bulbs. Short of killing him, they couldn't put a halt to Maxim's genius, so they did the next best thing: they employed him. As Edward Hewitt, an ex-employee of Maxim's later wrote:

> *In desperation they finally got together and made joint deal with Maxim. The terms were extraordinary. Maxim was to go to Europe for ten years, on a salary of $20,000 per year [approx. £360,000p.a. today], remain in close touch with all new European electrical inventions for the companies, but under no conditions was he to make any new electrical inventions himself during that time.*

Of course, what Maxim did was of little interest to the United States Electrical Lighting Company, they simply wanted him well out of the way. He was effectively being sent into an extremely comfortable exile. Suddenly he was faced with the thrilling prospect of being of independent means, and entirely free to pursue whatever course he wished. But what was he to do, for there was an entire world of possibilities in front of him? Maxim's mind was inextricably linked to the mechanical world, and the most advanced place in which to study either science or mechanics in the latter part of the 19th century was Europe. So, on 14 August 1881, he embarked on the SS *Germanic* for Europe, and arrived in Paris just in time to attend the Great Electrical Exhibition and accept the award of the Légion d'Honneur for his invention of the current regulator. While staying there, he began to give serious thought to what enterprise he wanted to devote his considerable energies to, but as is so often the case, the decision was made by a chance encounter in Paris with an old friend from New York, with whom he discussed his quandary. His friend's unequivocal response was forthright: 'Hang your chemistry and electricity! If you want to make a pile of money, invent something that will enable these Europeans to cut each other's throats with greater facility.'

That chance meeting was to have a profound effect not only on the development of firearms, but arguably the history of the modern world. Maxim realised that reviving his dormant interest in designing a working machine gun would actually serve several purposes. It would give him a focus for his limitless energy, provide the world with a defensive weapon that could perhaps

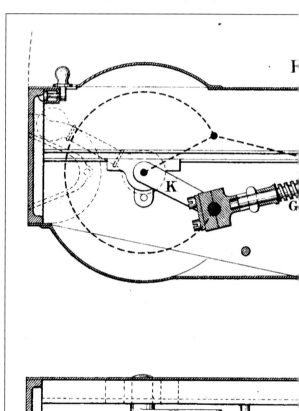

RIGHT Enfield Arsenal, 1910. Four Army artificers stand behind a two-barrelled Gardner, an 1893 rifle-calibre Maxim and a four-barrelled Nordenfelt. The Maxim appears sleek and modern compared to the other guns, although all three were still in service. *(Author)*

BELOW The complex breech mechanism of Maxim's prototype gun. The rotary feed drum is underneath the breech, and a series of rods and cams work to push back the rotating crank. *(Peter Smithurst)*

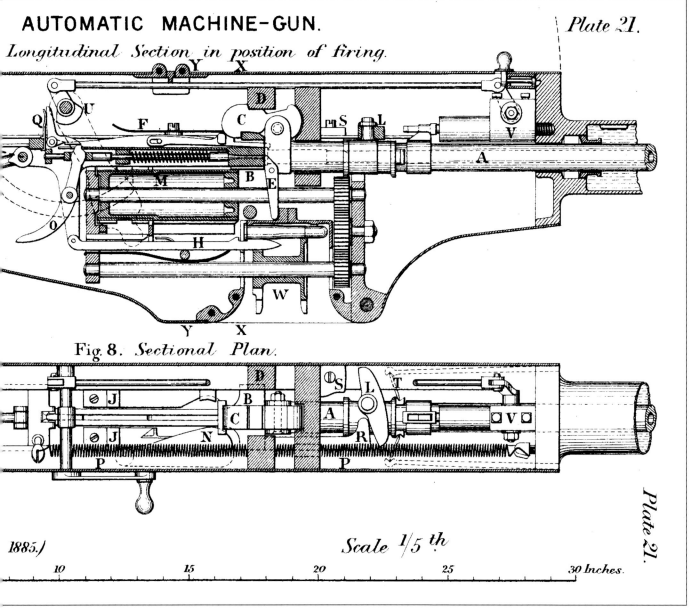

render war redundant and maybe even return him a profit on the expenditure and time that would surely be spent in its development. He worked quickly and had sketched out a rough diagram of a blowback, internally drum-fed magazine rifle by the time he left Paris. He then went to Cannon Street in London where he had established the Maxim-Weston Company to look after his financial assets, which it was conspicuously failing to do. While attempting to unravel the financial mess that the company had lapsed into, he read everything he could lay his hands on that related to machine guns, in particular the monthly magazine *The Engineer*, which he commented 'was without doubt the most reliable of all engineering publications in Europe'. He was particularly interested in the detailed patents and schematic drawings of the various automatic weapons that had been adopted for military service. During his address to the Institution of Mechanical Engineers in London in 1885, he had given a very shrewd overview of the shortcomings of what was available, which is worth repeating here in précis:

BELOW A scale drawing of the complete gun, with the traversing and elevation mechanism and the clock-hand firing regulator on the side of the receiver.
(Peter Smithurst)

> *The first practical machine gun is that made by Dr Gatling. The next machine gun of any note was that of Hotchkiss, another American inventor, who took it to France. This gun was followed by the Nordenfelt and Gardner. All four of these machine guns depend upon hand power for performing the various operations of loading, firing, and extracting empty shells. As considerable force is required for working either a crank or a lever, the gun has to be mounted on a very firm stand or base. This necessity precludes the possibility of turning [traversing] these guns with any degree of freedom: excepting the Hotchkiss gun, which is essentially a slow-firing gun [i.e., a cannon] firing only about forty shots a minute. These guns are each provided with a magazine of ammunition placed on top of the gun, and with any great rapidity of firing has, of course, to be replenished very often for which purpose two men at least are required, who are compelled to expose themselves above the gun [. . .] presenting a target to the enemy's fire. Their weak point does not lie here but arises from another cause. A certain percentage of all cartridges fail to explode promptly at the instant of being struck [. . .] they 'hang fire'. With cartridges over two years old it is not safe to fire a five-barrelled hand-operated gun over 300 rounds per minute, and even at this slow speed they will sometimes jam.*

Maxim was determined to overcome these issues, so typically looked to finding an entirely new means of solving them by designing a radically different system of operation, harnessing the recoil from the discharge by producing an improved mechanism that was, as he neatly expressed it, 'ammunition instigated'.

Encouraged by an unexpected offer from a group of London investors, he soon modified a Model 1873 Winchester rifle by fitting it with a self-cocking device that had a sprung butt-plate linked to a series of internal springs, effectively creating a semi-automatic rifle on which, as he pointed out, the shooter had nothing to do other than keep the trigger pulled. Unsatisfied by this, he determined to make better use of the recoil through harnessing the rearward force generated by the propellant, which was, as he wryly wrote, 'at best useless, at worst dangerous'.

One of his major problems was in finding a means by which the huge rearward pressure generated by a fired cartridge could be safely harnessed to move the breech-block. The continual problem he faced was that the pressure build-up blew out the base of the cartridge (the same problem Gatling had), thus releasing the gas. He looked again at his father's idea of a star-shaped feed-wheel, which

BELOW The rotary cartridge feed system of the prototype. It proved too fragile and complex for sustained use. *(Peter Smithurst)*

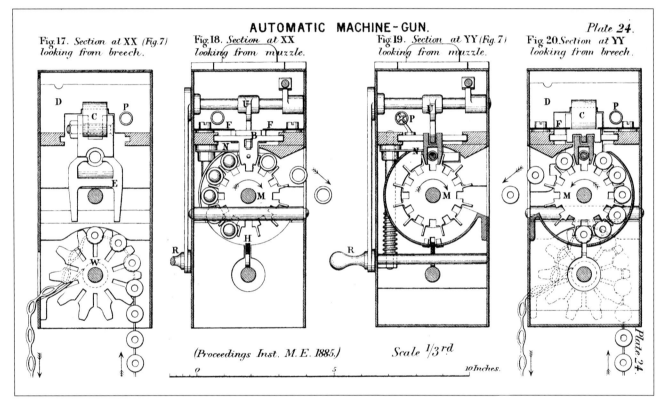

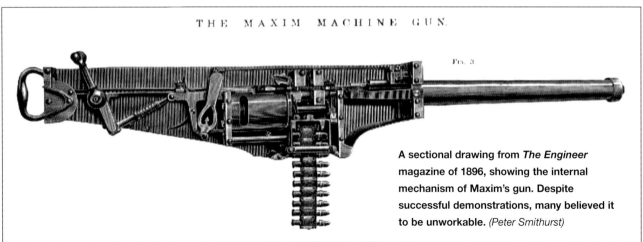

A sectional drawing from *The Engineer* magazine of 1896, showing the internal mechanism of Maxim's gun. Despite successful demonstrations, many believed it to be unworkable. *(Peter Smithurst)*

ABOVE A previously unpublished publicity photo of two of Maxim's employees test-firing a .303-calibre M1893 at 600yd. Every gun was tested prior to shipment. *(Author)*

BELOW Another rare image of a factory gun being test-fired to check the mechanism. This is on a fixed pintle mount at a range near Erith. *(Author)*

through an ingenious system of adjustable steel bars and a barrel and breech-block, managed to fire all six cartridges in half a second. He tried to replicate this feat, and to achieve this, he fired somewhere in the region of 1,000 .450-calibre Boxer-primed Martini-Henry cartridges for his test work, which proved a constant challenge. They were of commercial manufacture and variable quality. He noted that some came half-loaded and some were totally empty. Even those that worked were erratic in detonating, the bases tore off and pressure levels varied wildly. So frustrated was he, that he designed his own cartridge which proved crucial, for with reliable ammunition he was able to calculate that it required a pressure of about 60ft/lb to push a barrel back in, sufficient to operate an effective recoiling mechanism with a self-contained cartridge.

Working 14 hours a day, Maxim then tested every form of recoil operation he could devise, concluding that the most efficient way to make it work effectively was to 'utilise the force of the gases which issue from the muzzle of the gun at each discharge [. . .] to operate the breech mechanism of the gun, or to store up energy [. . .] for extracting the empty cartridge case, cocking the hammer, bringing another cartridge into position for firing [. . .] or preparing the arm for the next discharge'. He produced a blowback gun called the 'Forerunner' and while it never went any further than a single prototype, it enabled him to file 11 vital patents, including Patent 606 on 3 January 1884, which covered recoil operation. This is now regarded as a watershed in machine gun development, outlining the effect of the dynamic force of the muzzle-gas on a rearward-acting piston that operated as both a cocking and loading device.

He could not make enough of his own ammunition for his experiments, so he began to use a new solid-drawn brass .45-calibre cartridge case which was the current military issue for Gatling and Gardner guns. Because of the thousands he needed, he applied to the British Government for the supply of extremely large quantities; this elicited a puzzled request for an explanation, which Maxim provided. In an uncharacteristically benevolent gesture, the government agreed to supply him with as much as he required, free of charge.

He had by now set up a very well-equipped machine workshop at 57D Hatton Garden in London, with American-made machine tools shipped over specially, as he did not regard British-made lathes and milling machines as capable of producing the precision he required. Nevertheless there were still many problems to overcome. One of the greatest was in ensuring that his single-barrelled design did not suffer from the overheating that had plagued all other similar designs, and he looked again at Gatling's proposed idea of water cooling to provide an answer.

Maxim had calculated that firing a single cartridge would raise the temperature of 1lb of water by 1.5°F (the equivalent heat production would be capable of melting a 5lb ingot of iron). So providing a jacket containing 5pt of water would, theoretically, cope with a gun that could shoot at up to 600r/pm, provided this was not sustained fire. Having dealt with heat, he also had to consider two further fundamental problems, namely how he could ensure a reliable cocking and firing action and guarantee a consistent ammunition supply. Using the recoil motion of the barrel was already established in his mind, but he needed to speed up the rearward movement of the breech-block, so he designed his 'double-action accelerator', a piston that sat above the barrel and magnified the rearward motion of the block. He fitted this into a bell-crank breech, with a toggle-joint attached via a connecting rod to the breech-block. He realised this system had the double benefit of being immensely strong when locked, but also relatively easy to unlock, and his toggle rotated in an alternate circular motion: clockwise with one shot, anti-clockwise with the next – in much the same manner as a tolling church bell – and as a result the receiver housing was both large and heavy.

The final hurdle was that of ammunition supply. He had tentatively tried a simple looped canvas cartridge belt based on the issue US Mills infantry pattern, from which cartridges were fed via a rotating toothed wheel into an opening on the lower right side of the receiver. This was drawn in via a double-geared feed mechanism, but as the gun stood over a yard off the ground, and a loaded belt was extremely heavy, it placed a great strain on the feed mechanism, particularly the gears. Maxim toyed briefly with a drum magazine-feed system, but abandoned it. This statement actually belies the sheer amount of work that went into such a small side-project, for Maxim not only had to draw up plans, have a prototype produced with each component meticulously hand-made, test it and then be forced to accept that weeks of work were wasted when it failed to work satisfactorily. (Ironically, his drum design was later resurrected and used successfully in

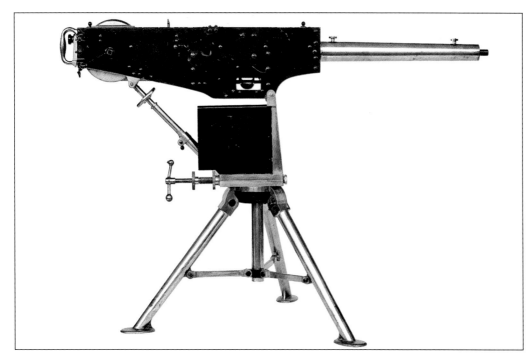

LEFT The prototype Maxim, one of two displayed at the International Inventions Exhibition in London in 1885. *(Courtesy Trustees of Royal Armouries Museum)*

RIGHT Hiram Maxim sits on a Maxim-Nordenfelt gun that has been fitted with an armoured tripod. This was taken at Livermore Falls, Maine, during one of his sales trips. *(Author)*

the Lewis gun, and Vickers Gas-Operated aircraft gun.) He returned to the belt concept, believing it to be the only practical solution. In an attempt to reduce the problems associated with the black powder-loaded .45-calibre ammunition, he also tried to lubricate the barrel and purify the powder residue using liquid. Where this might have taken him is anyone's guess, but happily for Maxim, 1885 saw the introduction of the new nitrocellulose powder, which solved his immediate problems, although it created others.

So far all of the calculations for his mechanical actions had been based on the pressures and heat generated by black powder. The new smokeless powder burned faster and produced far higher chamber pressures, so he had quite literally to go back to the drawing board and recalculate the stresses placed on every single component.

His prototype machine gun was finished in time to be exhibited at the International Inventions Exhibition at Crystal Palace in April 1885, where it won a Gold Medal. The gun was huge, 62in long, weighing 85lb and standing 3ft 6in high on a heavy tripod. As was typical of the time, it was beautifully made of blued steel and gunmetal, with a large brass cocking handle on the right side of the receiver and curved rate-of-fire dial above and behind the feed slot. This had an elegant pivoted adjuster arm that could moderate the firing speed from between 1 to 600r/pm. It worked well enough, but Maxim knew that to be a commercial proposition it must be simpler, lighter and cheaper to manufacture.

Never one to rest on his laurels, he immediately returned to his workshop, eliminating the gears, accelerator and rotating crank. He redesigned the crankshaft with a shortened toggle-joint that broke upward instead of rotating and introduced a simple but brilliant fusee spring that sat on the exterior of the left receiver plate. This was effectively an energy-storing system, attached to a cam worked by a short length of bicycle chain. As the gun fired and the breech-block moved backwards, the spring stretched but at the limit of its travel the energy it contained overcame the rearward inertia of the block and pulled it sharply forwards, returning it to battery, which placed the feed-block in the correct firing position. He described the major components now as comprising a 'barrel, side-frames, and crankshaft-operated breech block, all mounted in a non-recoiling outer frame [the receiver] and free to recoil to and fro therein with each discharge of the weapon'.

The one solution that still seemed to elude Maxim was in perfecting a system that not only fed the cartridges into the barrel, but also extracted the fired cases and disposed of them in one movement. He abandoned the feed-wheel and took a totally different approach to the problem by using a vertically sliding cartridge carrier (often referred to as the extractor) located in machined grooves on the lock-face. To function, it was necessary to cock the gun twice, which on this model meant pushing the handle forwards twice. This permitted the carrier to fulfil two functions. On the first movement it collected a cartridge from the belt in the feedblock, and as the mechanism moved to the rear the carrier slid downward on the lock-face. As the lock moved forwards again, the cartridge gripped in the jaws of the carrier was inserted into the chamber. The second cocking action then pulled the lock and carrier to the rear, the carrier rising upwards as it did so, sliding on the internal machined guide rails of the sideplates, grasping another cartridge from the belt in the feedblock. The firing mechanism was now cocked and locked forwards and would remain there until the trigger was pressed. When the gun fired, gas recoil forces took over, forcing back the lock and moving the barrel rearwards by ⅞in, as well as moving the carrier upwards to repeat the cycle again. The genius of Maxim's design was that this rearward motion also permitted the carrier to extract the fired case from the breech, and eject it from the floorplate of the receiver.

The new gun was patented in the summer of 1885 and named the 'Transitional Model'. It bore all of the hallmarks of the now-familiar Vickers gun, albeit initially having a pistol grip and trigger, but it soon acquired more comfortable and controllable twin 'spade' grips, a cocking handle on the right side and a new feedblock moved from below the barrel to above, with the gear wheels eliminated. Maxim had listened with great attention to the advice given to him by Sir Andrew Clark, Britain's Inspector General of Fortifications, and a very experienced soldier. Clark told him flatly: 'Do not give up until you make it so simple that it can be taken apart, examined, and cleaned with no other instruments but the hands.' Ammunition was held in 333-round canvas belts, supplied from a 'box or hopper' as Maxim termed them, which mounted on the side of the gun, thus reducing the drag on the feedblock.

ABOVE Maxim's second Transitional Model of 1885. Slimmer and lighter, it featured a simpler crank-locked breech mechanism and had both a spade and a pistol grip, though the latter was soon dispensed with. *(Courtesy Trustees of Royal Armouries Museum)*

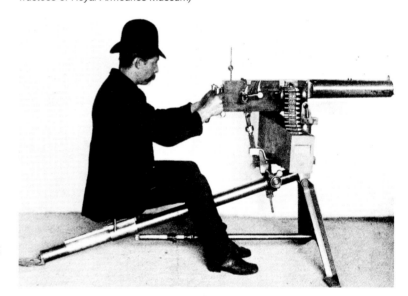

CENTRE Maxim's British assistant, Louis Silverman, seated on the tripod of a Transitional Model, showing its centrally mounted ammunition box and impractically heavy tripod design. *(Author)*

RIGHT Maxim behind his first production 'World Standard Gun', 1887. Note the steel shield and ejected cases in front of the tripod. *(Author)*

Chapter Three

The Vickers connection

Trying to manufacture his gun brought Maxim to the edge of bankruptcy, and he was saved by joining forces with his greatest rival in the firearms world, the Nordenfelt Company. He rarely did anything predictable and subsequently teamed up with the Vickers Company, a shipbuilding concern who had no experience whatsoever of small-arms manufacture.

OPPOSITE The first .303-calibre Vickers Service Model 1904 gun. It is on a Mk 'J' tripod set up to fire over a parapet. The gun displays the early five-arch top-cover, Maxim pattern rear sight and the fluted barrel adopted for this model. It also has the new 1904 pattern muzzle attachment, later to become standard on all Vickers guns. *(Vickers plc)*

ABOVE Maxims made few air-cooled guns, this example being the 'Extra Light Rifle Calibre Gun', weighing 44½lb with tripod. It was the only one made with the fusee spring inside the receiver. *(Courtesy Trustees of Royal Armouries Museum)*

The original Maxim workshop in Hatton Garden was incapable of manufacturing guns on a large scale and considerably more funding was needed to help in development. So, on 5 November 1885, a new partnership was formed with the long-established firm of Vickers, Sons and Company Limited. This was in many ways a strange alliance, for Vickers were large-scale manufacturers of armour plate for the Navy, and had no practical experience whatsoever in the manufacture of small arms. A more pragmatic choice would have been W.G. Armstrong and Co., ordnance and artillery ammunition manufacturers, but the problem was that they already had a reciprocal agreement with the Gatling Company and could not manufacture on behalf of a rival company. The agreement signed with Vickers made Maxim a director of a new 'Maxim Gun Company', but more importantly injected £50,000 in investment funds into the new venture (around £1.3 million by today's standards).

Permitting himself a little light relief, Maxim had taken time off from his work on the Transitional Model to develop a large-calibre gun, the Quick-Firing One-Pounder Autocannon, later to be known as the 'Pom-Pom', from the noise made when it fired. The technical details need not concern us greatly here, other than to say that this 37mm naval gun (not by accident was it identical in calibre to that of the Hotchkiss) employed all of the characteristics of the new Transitional Model but on steroids. It could fire at up to 300r/pm, weighed over 400lb without a mount and had a range in excess of 4,500yd. Albert Vickers enthused about the new gun, writing that: 'I would not sell it out [*ie* sell the patent] for less than quarter of a million pounds, and if our Government has any brains, they would pay that price to keep it secret.'

Eventually, Maxim had to return to the knotty problem of trying to perfect his Transitional Model; despite the new crank, the lock with its sliding carrier and fusee spring all operating to transform the working of the gun, he was still unsatisfied. In part this was because in 1886 the British Government had

BELOW One of the earliest 'Perfect Guns' on a wheeled carriage. Its beautiful construction with polished brass grips and ammunition tray, rolled feedblock with mahogany roller all represented the pinnacle of Victorian engineering. *(Courtesy Trustees of Royal Armouries Museum)*

finally decided on specifications for a service machine gun. It should be: 'Of a maximum weight of 100lbs, must be capable of firing 400 shots in one minute, 600 shots in two minutes and 1,000 shots in four minutes, pass all sand and rust tests and be capable of being field-stripped without recourse to special tools.' It was thus fortunate that Maxim's latest incarnation, 'The Perfect Gun', happily met all of these requirements. It weighed 60lb, was capable of firing in excess of 600r/pm, its top and lower covers prevented the ingress of dirt and it was mounted on a small-wheeled, lightweight carriage that incorporated a holder for the ammunition box. It was still chambered for the .45-calibre cartridge, although not for much longer.

Maxim was invited to demonstrate the gun at Woolwich Arsenal and took the latest example of the gun, now pared down to a winsome 40lb and mounted on a small brass naval plinth. This he had cleverly modified to act as a water reservoir. He enlisted the aid of a belt-loader, whose job was not so much to feed the ammunition into the gun (it did this automatically) but to ensure that the fast-moving belt did not become twisted and cause a jam. Maxim fired two 333-round belts, noting 'Most of the bystanders had to retire before the belt was finished on account of the dreadful fatigue to the ears.' The test was a triumph, and as a result he received his first British Governmental order, the three guns he demonstrated being purchased on the spot for trials.

This was just the beginning, however, for around the rest of Europe rising political tensions resulted in renewed interest over the potential of machine guns, and in 1887 he was invited to Switzerland for a machine gun trial against his other competitors: Gardner, Gatling and Nordenfelt. This could prove to be a make-or-break situation for the fledgling company but the Maxim gun proved equal to the task; firing at over 600r/pm, it shot the centre out of the target at 200m (the Gardner made a series of indiscriminate holes) and at 500m the Maxim achieved much the same result, but it was then required to be fired at 1,200m (1,312yd). This posed a considerable problem, for the rear-sight was only graduated to 1,000yd. Maxim calculated where the sight should be set and adjusted the traverse on the mount to just cover what he admitted was merely a 'blue streak on the horizon' (the target was a dummy battery of blue-uniformed artillery) and he commenced firing.

Shooting in short bursts, and traversing the gun gently to cover the target, he waited anxiously to find out the result. A message arrived saying that 'Technically we had killed three-quarters of the men and horses.' This was undoubtedly the first example of long-range harassing fire and after another demonstration in Italy, where it beat the Nordenfelt in every test (which included three days' immersion in seawater and firing without cleaning it), the Maxim reinforced its position as the premier machine gun then in production.

Significant orders soon followed and an Italian military order for 26 guns was his first official purchase by a standing army. This did not sit happily with the Austrians, as enmity between the two countries stretched back for decades. They were on the point of ordering Nordenfelts when, on 21 November, Maxim wrote and begged to be allowed to demonstrate his Perfect Guns. After successful firing trials, including his now well-practised trick of 1,000m shooting, the Austrians ordered 161 guns.

A significant event occurred just afterwards, when Maxim was approached by His Royal Highness Prince Esterhazy, who looked

BELOW A frequently reproduced 1896 publicity photo showing, from left rear: Albert Vickers, Sigmund Loewe and Li Hung-Chang, the most powerful statesman in China, with his assistant standing next to Hiram Maxim. The Perfect Gun was much used for felling trees. *(Vickers Archive)*

RIGHT A .45-calibre Maxim-Nordenfelt of around 1889 on a rare 'Carriage, Parapet, Machine Gun, Maxim Mk III' mounting, enabling it to be raised up to fire over breastworks. This gun was subsequently converted to .303 calibre. *(Courtesy Trustees of Royal Armouries Museum)*

BELOW The Maxim Gun Company's works at Crayford, c1895. Several Model 1893 rifle-calibre guns sit on their tripods, and on the right is an 'Acland' pintle mount for naval use. *(Vickers Archive)*

thoughtfully at the guns, and in response to Maxim's enquiry if he had fired fast enough for his liking, shook his head in sorrow and said: 'Ah, indeed, too fast. It is the most dreadful instrument that I have ever seen or imagined.'

By now Maxims were eclipsing all other rivals and Thorsten Nordenfelt saw the writing on the wall as far as future sales of his own weapons were concerned. Nordenfelt only made small numbers of guns, but supplied them to some of the most wealthy and high-profile people in the world. Thus, he had the contacts, as well as a 10-acre manufacturing plant at Erith in Kent, but not the orders he needed to remain solvent. On the other hand, Maxim had the guns everyone wanted, but not the manufacturing capability or sales backup that he required. Nordenfelt sensibly approached Maxim with an offer of a merger and a shares package, which was accepted. Maxim was moving on.

The year 1888 was a momentous one for Maxim. In July the Maxim-Nordenfelt Guns and Ammunition Company was officially incorporated. Maxim and Vickers had thus acquired even

greater financial backing, including £1.5 million in investment, a truly colossal amount at the time equal to £158 million nowadays. In addition, Maxim was to have the larger workforce at Erith. Even this was still inadequate to meet the growing clamour for his guns, so a 2½-acre site was purchased at Crayford in Kent, to be the new 'machine-gun manufactory'.

Maxim was now firmly on the road to financial success; his gun had beaten the opposition hands down. None could match its range, reliability or rate of fire, and one might be forgiven for thinking that he would relax a little and enjoy the fruits of his labour. Of course he did not, for science was advancing at an almost indecent pace and Maxim had to keep abreast with it.

Ammunition still posed a particular problem, for his guns had all been designed for black-powder cartridges and the large calibres used (usually .450in, 10mm or 11mm) merely required the gun to be supplied with the correct-sized barrel and carrier to hold them. The fusee spring to control the recoil was fully adjustable and the functioning of other components remained the same because the low pressures generated by black powder made little difference to the efficiency of the guns. This all changed dramatically with the adoption by the British Army in 1891 of the new smokeless .303in cartridge. At the same time so too did most of his European customers. The Swiss chose 7.5mm, the Italians 10.4mm and the Austrians 8mm, which meant that each gun now had to be reworked to enable it to function properly with the varied chamber pressures generated by the smaller smokeless cartridges. Maxim was forced to recalculate the energy required to work the crank mechanism and breech-block to cope with the longer pressure-curve of these progressive powders.

He began to incorporate all of his ideas into one fully formed gun, a new 'World Standard',

as he termed it, which had first appeared in 1887 in .45 calibre. These were offered as a complete package comprising gun, tripod shield, elevation and traversing gears, spare barrels and lock, parts kit and 10,000 rounds of belted ammunition in any calibre at a price of £3.00 per thousand. The price was £250 per gun, which was not cheap (equating to around £26,000 today) but despite this, demand continued to increase.

A look at the surviving order books shows substantial orders from the British War Office,

ABOVE The Nordenfelt works at Erith where not only the rifle-calibre guns were made but also the 1lb 'Pom-Pom', visible in the foreground. *(Vickers Archive)*

RIGHT A Model 1908 Light gun on its wheeled, armoured mount. The unpopular Peddie-Calochiopulo rearsight can be seen on the top cover behind the water jacket. Some 268 of these guns were shipped to Russia, where they and their wheeled armoured mounts evolved into the Russian M1910. *(Author)*

ABOVE The Maxim M1893 'Rifle Calibre, .303 inch, Machine Gun' was the most widely used in the British Army prior to the introduction of the Vickers. *(Courtesy Trustees of Royal Armouries Museum)*

Spain, France, Holland, Italy and Argentina, as well as assorted Crown Colonies in India and Africa. These guns now bore all the hallmarks of the Maxim models, with handsome gunmetal water jackets, deeply blued receivers, brass fusee covers and twin brass spade-grips. A simpler steel tripod had also been developed that did away with the rather heavy gun carriage, although this was still an optional extra and it usefully incorporated a seat bolted to the rear leg for the gunner. The rate of fire had been slowed to around 450–500r/pm.

In a surprise move, in the middle of 1887, the German Kaiser Wilhelm I asked for a demonstration at Spandau Arsenal. The Germans wanted a gun to fire the old black-powder 11mm Mauser cartridge (they did not adopt the new 7.92mm cartridge until the following year). Ensuring the guns could shoot these was not a problem, but Maxim had no stocks of cartridges to hand, nor could he find any, so had to assemble an emergency workforce to cast and hand-load sufficient cartridges.

His gun was to be pitted against Gatling, Gardner and Nordenfelt guns, and while all of the guns each fired 333 test rounds successfully, examination of the targets showed that the groups from the rival guns were very scattered – some missing completely – while the Maxim with its rigidly supported gun, shot the centre of the target out. The Kaiser pointed at the target and said 'This is the gun . . . there is no other.'

It was during this demonstration that the automatic traverse of the gun was accidentally set, turning it towards the entire German General Staff as it fired. Within seconds they would have become the Maxim's first and highest-profile victims, but he leapt forwards and disconnected it. Thereafter, the auto-traverse was no longer supplied except by special order.

Although he did not realise it at the time, this was to prove a fateful demonstration for two reasons. First, it led to a seven-year agreement with the armament giant Ludwig Loewe of Berlin for the licensed manufacture of Maxim guns and, second, it was also the initial step in the separation of manufacture that was to lead to two variants of the same gun, the Vickers-Maxim and the German Maxim being produced. While the former was to be extensively modified and improved as the 'Class C' Vickers in British service, the latter remained essentially the Model 1894, which would metamorphose into the now familiar Maschinengewehr 1908 (MG08).

Despite his remarkable abilities, Maxim could not work efficiently as engineer, designer and business manager, and the financial affairs of the new company could best be termed as wayward. From 1886 their holdings had expanded to incorporate both new factories. Erith employed 400 men and women, Crayford 750, and facilities in Dartford and Birmingham with 180 employees existed solely for testing and ammunition production. There were also range facilities for trials and a subsidiary company formed in Argentina to assemble component parts into finished guns for the South American market. But Maxim had overstretched himself and the company was showing unsustainable losses. That being the case, he offered Ludwig Loewe's younger brother Sigmund the job of overall manager, which proved to be a brilliant move. Within a year, Loewe had turned the loss-making concern into a business showing a profit in 1896 of £138,000 (now about £14.5 million). This was a turning point for both Maxim and the company; for the first time his guns were being manufactured outside Great Britain, his financial affairs were in order and he had the freedom to continue development work on his guns.

There was, however, a fly in the ointment. Vickers and Maxim were heavily involved in lucrative commercial sales, while still needing to meet the British Government orders. The manufacture of Maxims was particularly slow, almost a year behind schedule, and they

RIGHT Two Royal Navy sailors with a .45-calibre gun on a naval landing carriage, 1895. *(Author)*

CENTRE A service .303in Maxim on a model 1889 'Carriage, Field, Machine Gun, Cavalry or Infantry'. The impractical size and weight of the carriage ensured few saw action. *(Author)*

were proving much more expensive than the government was happy to accept. Furthermore, it was difficult for government inspectors to effectively quality control components.

In 1890 it was proposed that manufacture of all British Government Maxims be undertaken at the Royal Small Arms Factory at Enfield (RSAF). British military long arms had been made there since 1816, and the factory had been supplying very high-quality rifles exclusively for the Army since the introduction of the Pattern 1853 Enfield rifle-musket. Using Enfield would solve several problems: it provided the government with full control of production, ensured total component interchangeability (not always the case with Maxim-Nordenfelt-manufactured guns) reduced unit costs per gun and relieved Maxim and his associates of the burden of trying to simultaneously produce both commercial and military guns. Production began in 1891 and the first known gun still survives, dated 1892. For the next 23 years Maxims were made at Enfield, in both .45in and then .303in calibres, production ceasing only when they were declared obsolete in 1917, after 2,568 had been produced (with the Maxim-Nordenfelt Company receiving a royalty of £25 per gun).

At this time, however, Maxim's dominance in the field of machine guns was being challenged; the brilliant American engineer and gunmaker John Moses Browning and his brother Matthew had devised an air-cooled gun that was put into production as the Colt 'Gas Hammer' machine gun. The story of the competition between the two companies for the lucrative American market is too long to repeat, suffice to say that despite turning his attention to improving and modifying the World Standard gun to produce a much lighter variant (the 28lb Extra-Light Rifle calibre gun), the US Navy spurned Maxim and in 1895 adopted the Colt instead. Far from deterring him, it merely spurred him on to try

BELOW The very first Enfield-made '.45 Gatling Calibre Gun, No 1' on its 85lb 'Tripod, Mk I'. The tripod was too heavy and never officially adopted for service. *(Courtesy Trustees of Royal Armouries Museum)*

RIGHT A British officer test-fires a .303-calibre British issue M1893 Maxim gun during the Boer War. *(Author)*

to capture his own home-country market and the work to produce an acceptable American variant will be looked at in Chapter 7.

Closer to home, there were important changes afoot for the old Maxim Gun Company, and its alliance with Nordenfelt who were by then well established as manufacturing partners. Vickers were in an odd position, providing investment but still no actual manufacturing capability. This was put into perspective in Vickers' own history: 'It had become clear that some years before 1897, the company [Vickers] were occupying a half-way house. They were still essentially a steel firm, with one foot in armaments.' In contrast, Nordenfelt were experienced gunmakers by this time, with a skilled workforce capable of making the Maxims to the required standards. Indeed,

RIGHT A British service Maxim being fired from behind a *sangar* (a stone-built protective shelter) during the Second Boer War. It is interesting to note that it appears to have a leather cover over the water jacket. *(Author)*

this was acknowledged, at least up to the end of 1897, by the fact that fusee covers of all guns were marked with the Maxim-Nordenfelt name. However, the Vickers brothers, Albert and Tom, were determined to diversify into light armaments, so entered into a contract with Maxim on 1 October 1897 to rename the company Vickers Sons and Maxim (VSM). It would not be a rapid change, as tooling-up for such a complex product could take up to two years, but Albert was confident, subsequently writing to the War Office: 'Provided we are allowed a sufficient period of time, we are prepared to manufacture guns and mountings up to any amount.' This amalgamation was also to mark the transition from Maxim being in sole charge of the manufacture and development of his guns and chairman of his own company, to the eventual severance of his ties with the Maxim company and its products, and the rise of Vickers as the country's premier producer of machine guns.

Maxim's interest in his guns was waning; he believed he had developed them as far as was practical and his ever-restless quest for new challenges was leading him, among other interests, into aviation. He became a British subject in 1900, was knighted in 1901 by Queen Victoria and the following year sold almost his entire shareholding to Vickers, signing an agreement to cease any further work on the guns that bore his name, in return for a generous director's fee of £1,200 per year. He didn't need it, for he was already a multi-millionaire, but it was perhaps his last tangible link with the guns he had worked so tirelessly to perfect. Vickers retained all of Maxim's designs and patents along with the Erith and Crayford works and employees, as well as the irreplaceable Sigmund Loewe. But changes needed to be made, for Vickers engineers had taken a long, hard look at the World Standard guns and their component parts and come to the conclusion that they needed to be modified for the sake of manufacturing simplicity, functional reliability and cutting costs.

The old guns were things of Victorian engineering beauty, blued and polished, but all of this was time-consuming to manufacture and unnecessarily expensive. There were some mechanical issues with the early guns as well, and Vickers introduced a number of improvements in 1901. The most noticeable was the substitution of steel for brass on the water jacket; it was still smooth, but was lighter and easier to manufacture. The lock was modified, the old 1896 type still requiring tools to strip it, so a new pattern was drawn up that needed no tools for dismantling except a punch to push out the locking pins. More important perhaps was a modification that incorporated a simple but effective cartridge head-spacing arrangement using thin washers that fitted on the connecting rod crank. From now on any new lock could be

LEFT A grainy but interesting contemporary newspaper photo of the Natal Carbineers with a M1895 Extra-Light Maxim captured from Boer soldiers during the fighting for Lombard's Kop, 7 December 1889. *(Author)*

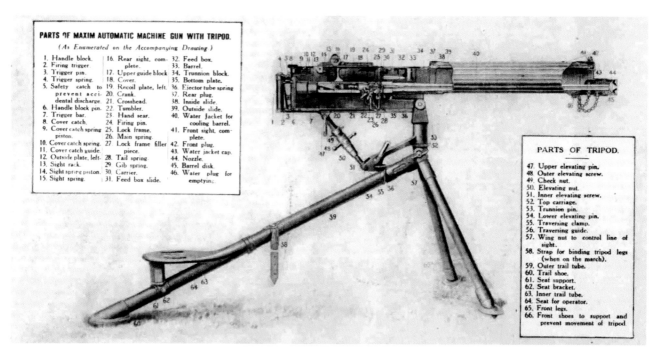

ABOVE The first published drawing of the Model 1906 'New Light' gun, from *The Engineer* magazine. *(Peter Smithurst)*

BELOW A New Model 1906 'New Light' Maxim, sporting its new fluted jacket, mounted on the VSM Mk 'B' Tripod. Probably the finest of all the Maxims made, this is the only known example. *(Courtesy Trustees of Royal Armouries Museum)*

field-stripped without the need for special tools and be accurately head-spaced with the spacers provided in the spares kit. Although all Vickers-Maxim guns post-1901 were supplied with two locks, it was now possible, unless damage was excessive, to field repair a broken lock.

The cessation of the seven-year agreement with Loewe in Berlin at this time meant that the Germans were now free to make their own guns, which were to be manufactured at the state arsenal, Deutsche Waffen- und Munitionsfabriken (DWM) in Berlin. Strangely, throughout their service life, the Germans never made any attempt to improve the original lock design, thus requiring each MG08 to have three dedicated locks that were not interchangeable with any other gun.

Vickers also reduced the size of the feedblock and dispensed with the distinctive wooden feed-roller. An improved-shape crank handle was fitted, as well as a host of smaller detail changes, but this was not by any means an end to the modifications.

They continued to look at ways to reduce the weight of the guns, so in 1906 a new model, the VSM 'New Light' Model was introduced. This was arguably the finest gun of genre, certainly in terms of quality of manufacture and at this time it finally acquired its unique corrugated water jacket, designed by Trevor Dawson and George Buckham of Vickers. Despite requiring specially rolled steel sheet, it provided both greater strength and an increased cooling area compared to the old smooth steel or brass patterns. Weight was also shaved off by using lighter high-grade steels wherever possible, reducing the 60ls of the brass 1893 model to 40½lb.

Probably the most significant addition was a new muzzle booster, developed by VSM, comprising a cupped steel disc that sat inside

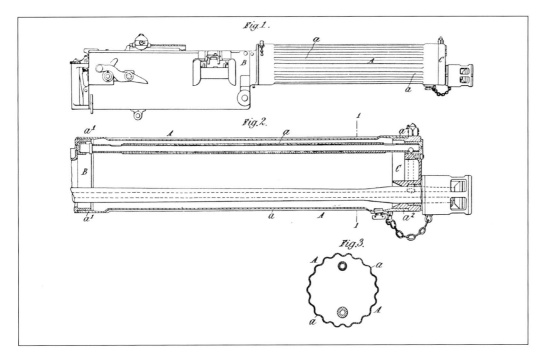

LEFT **The patent granted on 24 March 1906 to Trevor Dawson, RN, and George Buckham of Vickers, Sons and Maxim, for the new corrugated jacket design. It made the Vickers gun the most instantly identifiable machine gun of the era.** *(British Patent Office)*

a perforated sleeve that locked to the water jacket cap by means of a quick-release bayonet fitting. This harnessed a greater proportion of the propellant gas, forcing the barrel backwards faster and more positively which, allied to the lighter internal components, reduced inertia and sped up the rate of fire. It also captured a considerable amount of the fouling that would otherwise have lined the muzzle and barrel, giving it a longer life and reducing the amount of cleaning time. The new water jacket held a little more water, a shade over 7pt as opposed to 6, and in testing, Vickers determined that when fired in short bursts, 2,000 rounds could be expended before any water was needed to top up the barrel.

These guns were termed Type 'B' by Vickers. This was because Vickers believed that naming guns such as 'New Light' was no longer suitable and did not convey sufficient military gravitas upon them; therefore in keeping with their tradition of shipbuilding, it was decided by the directors that machine guns would be given a model designation, and not a name. As ships were manufactured according to their class, A Class or B Class destroyers for example, so this should be applied to the Vickers guns.

But deciding what to call their guns was the least of their problems, for the Germans were outselling the Vickers-produced guns by almost two to one. This was in part because they were cheaper to manufacture (DWM had massive small-arms manufacturing capabilities) and the Germans had a well-deserved reputation worldwide for the quality of its small-arms industry. Vickers needed to come up with new ideas to keep their guns in production, reduce still further their manufacturing time and costs and make themselves more efficient than DWM. Most importantly, they had to use their own patentable ideas to achieve this.

One factor that immediately stood out was the size and weight of the receiver body, which had to include the toggle-joint as it folded downwards on recoil. Dawson and Buckham wondered what would be the result of reversing the action on the toggle, to break upward instead, so they made a prototype that did exactly that and it worked exceptionally well. This development work took two years, but in 1908 the new gun was introduced. Confusingly for the soldiers trained on the old Maxims, this meant that the crank handle on the new guns now had to be pulled *backwards* twice, instead of forwards. In keeping with policy, the new machine gun was termed the Class 'C' Vickers, and brass plates riveted to the top cover proclaimed it to be the 'Vickers Automatic Gun RC Gun, Class C', 'RC' standing for Rifle Calibre. At this point the Maxim and Vickers stories diverge, as the old-model Maxims continued to be manufactured by Germany as the MG08, but the Vickers gun had now begun a long life of its own.

Chapter Four

First World War manufacture

In 1914 the British Army had one-third of the number of Maxim machine guns possessed by Germany and almost no production capability, yet by 1918 Vickers were out-producing Germany. This resulted in both Britain and Germany facing each other across No-Man's Land for the duration of the war, each using virtually identical versions of the same machine gun.

OPPOSITE The concept of motorised infantry was an alien one when the Vickers was first introduced, but the First World War introduced many innovations. One was the use of motor machine-gun units. Here an immaculately prepared Clyno outfit is being inspected at Deival, June 1918. *(IWM Q010325)*

ABOVE Mouquet Farm, Somme, summer 1916. Two Vickers gunners fire at a nearby target, while behind them another pair use a captured MG08. Every gunner was instructed in the use of German machine guns. The cautious demeanour of those observing indicates the enemy are not far away. *(Author)*

BELOW A 'five-arch' lightened top cover from a Vickers gun serial No L1053, made at Erith in June 1915. This time-consuming modification was omitted from later guns. *(Richard Fisher)*

There now existed two types of Maxims, the old models that were being manufactured in great numbers by DWM in Germany, and the British-made Vickers and Enfield guns. That the two main protagonists in the First World War should end up facing each other behind the sights of nearly identical machine guns is ironic in the extreme, but by now there were very distinct differences between the models. The MG08 is not the subject of this book, so a very brief outline of the differences will suffice. By 1912 the most obvious visual change was that for the first time the two guns now looked completely different.

The MG08 had a smooth steel water jacket with two vertical trunnions for mounting to the heavy four-legged *schlitten* (sledge mount), and a much deeper receiver. In contrast the Vickers had a plain jacket with a machined bottom mount for the locking pin, and a much slimmer receiver. The Maxim's cranking handle was higher up on the right side of the receiver and the spade-grips were larger. It retained its long, one-piece top cover, while on the other hand the Vickers had a split top cover, the rear giving access to the lock and toggle mechanism, the front permitting the feedblock to be removed without disturbing other components. The snub-nosed muzzle booster of the Vickers was markedly different to the long trumpet-shaped cone of the Maxim. This all made the MG08 considerably heavier, with gun and mount weighing in at a shade over 150lb, including water.

While improving the Vickers was a necessary requirement, changing the action of the toggle meant there was now virtually no room beneath the top cover, and this posed a problem for a rear sight mounting, which required clearance underneath so as not to foul the internal parts. A new ramp-type sight was designed that mounted into a pair of machined blocks at the rear of the top cover. This had a raised strengthening rib, which ran the length of the rear top cover and on early guns this had lightening slots cut out of it, leading it to resemble a five-arch bridge (this has become an instant method of identifying an early production Vickers). Lightening the feedblock had another positive effect on the reciprocating parts of the mechanism, for it resulted in a shortened crank connecting rod that could take greater pressure, and the shallower receiver plates, with their machined internal guide ribs that the cartridge carrier followed, were thought to be able to withstand greater torque. This in turn meant the engineers used thinner walled steel and made weight savings, aided by the fact that the depth of the receiver sideplates had now been reduced by 2in, eliminating the wasted space below. This proved to be a false economy, as we shall see.

Moving the trigger bar up required repositioning of the sear and spring and this produced the unintentional effect of enabling an experienced gunner to fire single shots. The feedblock was also improved, being lightened and having the internal pawls modified. These held the belt as it was fed through the mechanism, and were changed so that pressing on them with the thumb could more easily disengage the belt. A sliding dust cover was incorporated into the underside of the receiver, eliminating the possibility of dirt entering, while still leaving a clear path for the fired cases to eject.

The overall effect of this was to reduce the weight of the Model 1906 to 28lb, some 30% lighter than the original New Light Maxim and precisely half the weight of the old British service guns. The War Office were pleased with the result and ordered a batch of 26 guns for troop trials.

There was then the matter of tripods to consider. The new gun had to be man-portable so the old wheeled carriages had been phased out, and a number of different three-legged mounts had been constructed for the Maxim models, none of which proved totally ideal. Vickers produced several commercial tripods of their own. They devised a Type 'B' that incorporated a complex crank handle for height adjustment that raised or lowered the gun and, to allow for this, the legs were mounted in vertical slots. The small bicycle-type saddle was retained on the rear leg for the gunner and it was effective enough, but there were

TOP The left side of the Mk I Vickers used for the detail photos in this book. It is an 'L' series gun (serial No L3663), manufactured at Crayford at the end of February 1916, and has the name 'Jean' engraved on the top rear of the water jacket. At the end of the war it went to Australia, most probably with the unit that had been using it, and was used for training purposes. When too worn out to shoot, it was marked 'DP' and relegated to instructional use. *(Author)*

ABOVE The right side of 'Jean'. All the accessories are of First World War issue, even the ammunition, which was recovered from a cache of belted .303 found on the Somme in 2009! The water tin shows its original civilian red paint under the green. The barrel jacket orifice cork plug and thread protector for the water connection are visible dangling from the muzzle. *(Author)*

RIGHT Many of Britain's Allies adopted the Vickers, including Italy. These strikingly picturesque Italian soldiers are being given Vickers familiarisation on the Piave Front in summer 1918. *(IWM Q012887)*

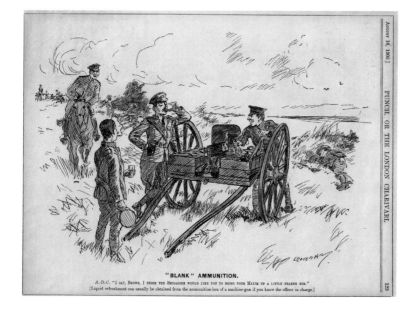

BELOW Not everyone took the new weapon seriously. Officers in charge of Maxim gun sections were regarded as being extremely fortunate, as this cartoon from a 1905 edition of *Punch* shows. *(Author)*

numerous problems with the design. Aside from the complexity of manufacture, the elevation gears were prone to jamming if fouled by dirt or mud and with its long rear tripod leg it was awkward to carry. It quickly went through a number of revisions, becoming in turn, the Models C, E and F, the latter two having the addition of a small wheeled mount that had an armoured shield. They were inevitably extremely expensive, and were sold only to Russia (see Chapter 7 for more on these guns). Otherwise, they were not a commercial success, so in response, Vickers brought out yet another model in 1913, the Model J, which eliminated the elevating gear and crank handle. Nevertheless these beautifully made commercial tripods were still too costly for large-scale military purchase, a fact brought sharply into focus when Italy ordered 890 guns from Vickers, without tripods, preferring to buy and adapt a cheaper Spanish design instead.

It was clear that an entirely new tripod was required that was simpler, lighter and cheaper and in January 1900 a Mk II tripod was introduced that replaced the old 85lb Mk I Maxim mount. This lightweight little item weighed a svelte 56lb, identical, in fact, to the weight of the old Maxim gun. Although certainly heading in the right direction, the Army deemed it still too heavy to be manageable, and they became exasperated to the extent that in 1905 the War Office had begun to design its own version, a simple tripod that had only a turn-wheel at the rear, connected to a vertical threaded shaft that provided the elevation. The body or socket of the tripod was cast in brass and was hollow to accept the tapered shaft of the crosshead that simply dropped down into it. The gun was held in place by two cross-pins (angled steel locking levers). The rear mount sat midway along the underside of the receiver, while the forward mount was at the very foremost part of the receiver, where it joined the water jacket. When the pins were removed, the gun could be lifted clear leaving the crosshead and tripod in situ. It also permitted a traverse of 180° unless stops were placed on to it. The legs were shorter than on the original service tripod, the front pair at 23in, the rear 25in, and the seat was dispensed with. The resulting mount was typically ponderously termed the 'Mounting, Tripod, .303 inch Machine Gun Mk IV'. It was introduced on 23 January 1906 and would remain, virtually unaltered, in service with the Vickers for the next 55 years.

British military interest in the new 'C' Class Vickers remained high. On 13 September 1910 a series of trials were organised at the Small Arms School at Hythe in Kent by the Small Arms Committee (SAC), using two standard production Vickers-manufactured 'C' Class guns equipped with Mk IV tripods and two of the older service M1893 Maxims. The trials were exhaustive and lengthy; after firing 2,000, 4,000, 5,000, 6,000, 8,000 and 10,000 rounds of .303in ammunition non-stop, every component was gauged and visually checked. Bullet velocities were measured, accuracy carefully noted, stoppages were logged and any failed component was set aside for later examination. Casts were then taken of the chambers in all of the used barrels at the end of the testing to determine wear. The testers tried their level best to make the guns malfunction, using heavily worn water-soaked belts having '. . . some cartridges being inserted ⅜inch further into the belt than others. Each gun negotiated the belt satisfactorily.' Dust and dirt tests were also undertaken, with sand being blown over the two Vickers guns. 'Both the Light Guns worked somewhat sluggishly, but did not stop firing. Fitting . . . the muzzle booster the guns worked as sweetly as usual without the sand being cleaned away.' Fusee springs were tightened to raise the rate of fire from 400r/pm to 650r/pm, placing a great strain on the internal components, but nothing of any importance failed. At the end of the testing, it was clear to all involved that the immense amount of work put into the new models had paid off, for they were better in every respect to the old guns.

There were hardly any mechanical failures noted apart from routine jams in the feedblock, often caused by the belt whipping as it was drawn through the mechanism, dislodging cartridges. There were no stoppages despite using a random mixture of new and old ammunition, one lock spring broke and was quickly replaced, and virtually no component showed any wear apart from a feedblock.

Using 250-round webbing belts the guns averaged 400r/pm and after 10,000 rounds had been fired, the loss of velocity was a negligible 200fps. The conclusion was unarguable, the testers noting that the Vickers scored highly for its lightness (it now weighed 37lb with a full water jacket as opposed to 67lb for the service Maxim. Indeed, the normally cautious SAC's summary was remarkably positive.

The new guns were '. . . far superior to the old guns, particularly as no tools were needed for the Vickers, whereas stripping a service gun a mallet and several punches are required'. The Vickers had '. . . considerable mechanical advantage over the Maxim and was superior in its ease of stripping and exchange of broken or worn components'. The wear measured on the internal components was in fact so small that it was not noted. Some minor recommendations were made to modify internal parts, and the biggest criticism was reserved for the 'Peddie-Calochiopulo' rear sight, which was poorly positioned too far forwards and only graduated to 2,000yd. The War Office wanted the sights graduated to 2,600yd, the same as the service Lee-Enfield, and they generally disliked the design. Vickers quickly responded with three new ones, and a simple tangent rear sight was selected.

The new sights and all the suggested internal modifications were to be fitted on to production guns and the 26 trials guns were all modified, to become known as the 'Gun, Machine, .303 Mk I', and were accepted by the War Office in the List of Changes No 16217 on 26 November 1912. The Vickers had finally been adopted into service with the British Army, and it was not a minute too soon.

By the spring of 1914, the complement of Maxim guns in British service stood at 1,846, being an eclectic mix of Maxims, Maxim-Nordenfelt, Enfield-Maxims and Vickers. Most were now chambered for the standard British .303 Mk VII cartridge, the first to be fitted with

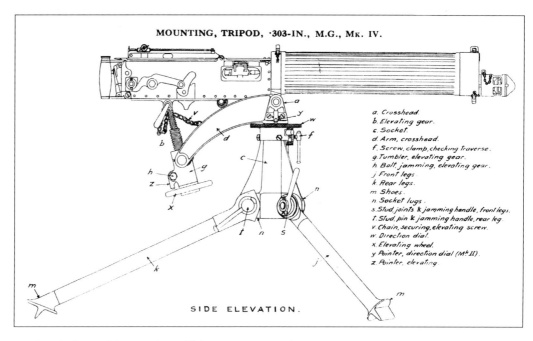

RIGHT The gun and parts drawing as supplied in the first Vickers handbooks, which were commercially produced, as no official manual was issued. *(Author)*

BELOW Most of the 12,000 employees at the Crayford works pose for the camera. The photo gives some idea of the colossal amount of manpower required to manufacture just one weapon during the First World War. *(Vickers Archive)*

a spitzer bullet and the new issue Vickers gun, the 'Gun Machine, Vickers .303 inch Mk I', although some Maxims in reserve service were still chambered for the old round-nosed .303in Mk VI cartridge.

Vickers, meanwhile, had been setting up production lines at Crayford and Erith to begin manufacture, but as of 6 August 1914, when war was declared, a mere 109 had been completed, all at Erith. It must be said that initially machine gun production was not regarded as a priority. After all, unlike the experience with the previous war against the Boers in South Africa, this war was expected to be short and sharp and no one had the slightest concept of the trench stalemate that was to become the primary characteristic of the fighting for almost three years.

A few units who mobilised early actually purchased their own guns, paid for by their commanding officers; the 1/4th Seaforth Highlanders ordered one with tripod, the 1/5th Gordon Highlanders two, and two even went to a mounted unit, the King Edward's Horse. The price of these guns was substantial, on average £165 each with tripod and spares (about £15,000 today).

But war notwithstanding, it had always been the War Office's intention to order more guns. Between August 1914 and June 1915, they placed four contracts with Vickers, initially for 192 guns, followed by a further 100. In September 1914 this was followed by two large contracts in quick succession for a total of a further 1,500 guns. The reason for this sudden increase was the very obvious fact that British forces were being outgunned at the front, each British infantry battalion having only two machine guns as opposed to the Germans' six, enabling them to field 4,422 MG08s by the time the war began. By 1915, depending on the unit, the Germans were able to increase this to between nine and twelve Maxims per battalion.

The problem facing David Lloyd George, then Minister of the newly established Ministry of Munitions, was in determining exactly how

LEFT A shortage of machine guns in the early days of the First World War resulted in an assortment of weapons being used in the field. Here a Maxim 1-pounder 'Pom-Pom' has been put to good use, judging by the number of empty cases. *(Lawrence Brown)*

many guns were required, bearing in mind the huge demands being placed on industry for every other type of munition. Meetings between the Deputy Minister Sir Eric Geddes and Lord Kitchener did not go well; on being asked how many Vickers and Lewis guns should be ordered, Kitchener exploded. 'Do you think I am God Almighty that I can tell you what is wanted nine months ahead?' In reality, this was precisely Kitchener's job, so when pressed by Geddes he gave a deliberately evasive reply: 'I want as much as both of you can produce.' Eventually when pressed to come up with a concrete figure he proposed two guns per battalion as a minimum and four as a maximum. Anything over four he stated was merely a luxury.

Lloyd George was unimpressed: 'That was the opinion of the Secretary for War, who was looked upon generally as our greatest soldier.' Lloyd George then spoke to as many field officers as he could on the subject and the general consensus was that it was simply impossible to have too many Vickers guns. He told Geddes 'Take Kitchener's maximum (four guns), square it, multiply that result by two; and when you are in sight of that, double it again for good luck.' Thus the figure that was arrived at, of 12,000 guns, was reached mostly by inspired guesswork. These were scheduled to be delivered at the rate of 50 per week and completed by mid-June 1916, which was fine in theory, but Vickers were already in trouble.

The company simply could not find sufficient skilled labour to begin to meet the demands of their *pre-war* contracts, let alone a new one for several thousands of guns. Many skilled workers had left their jobs in 1914 and enlisted, creating a manpower shortage that could not be quickly alleviated. Of course, it was possible to train up new workers, in particular the women who now began to flock to industry to replace the absent men, but this took time, and time was not on Vickers' side. It was further clear that both the factories at Crayford and Erith would have to be greatly expanded to try to meet the demand – Crayford, the largest facility, being given a target of 400 guns per week, and Erith 100. The thorny question was how to achieve this? Both sites had the space but not the workforce.

The post-war *History of the Ministry of Munitions* outlined the problem succinctly. 'The greatest check on the extension of the factories was the serious difficulty in securing adequate skilled labour for setting up the machine. In October 1915 . . . skilled machine setters who were released from the colours should be drafted to Crayford . . . [but] the numbers so secured were not adequate to provide the rapid installation of plant.' Within the factory attempts were made to produce skilled labour by encouraging skilled workers to train others. A bonus of £1 was offered to every apprentice who became proficient at setting up their own tooling.

ABOVE Workers finishing off and fitting sideplates at Crayford. *(IWM Q110353)*

BELOW A line of women machinists manufacturing the sideplates for the Vickers. *(IWM Q110354)*

BELOW RIGHT A female worker machining sideplates for the Vickers gun at Crayford. *(IWM Q110352)*

Of the original pre-1914 War Office order for 1,792 guns, only 1,022 had been delivered by July 1915. Manufacturing 12,000 for delivery by mid-1916 was clearly impossible. To muddy the waters still further, a second contract had been accepted by Vickers to supply the French Government 2,000 guns with commercial 'J'-pattern tripods. The War Office promptly and wisely prohibited any further commercial sales, and stipulated the first 48 of these guns should be manufactured at Erith in order to provide the workforce with much-needed experience; thereafter they would be manufactured at Crayford. In practice it took Crayford 6½ months to make the first 952 guns, but only 4 months to make the next 1,000. The French got their guns (which by all accounts worked very well), while the factory workforce got the experience it so desperately needed. But there was still no possibility of meeting the British Government's huge order.

In a forlorn attempt to try to alleviate the problem, Vickers entered into an unworkable agreement with the Colt's Patent Firearms Company of Hartford, Connecticut, to supply 6,000 of the required guns. This would have undoubtedly helped, if Colt were actually tooled-up for production, which they were not.

As the fighting on the Western Front intensified, so did the problems for Vickers, for not only were they unable to fulfil their original contractual obligations, but the British Army decided in January 1915 that the machine gun requirement for each battalion be increased to four guns. This did not take into account machine guns desperately needed by the Royal Flying Corps (RFC). Then there was the problem of wastage – Vickers guns lost or destroyed by enemy action – which by 1916 were calculated to be 50% per month of production, equating to around 800 guns that would be additionally required on top of those already ordered.

The urgency for production to begin can be glimpsed from the fact that despite not having the required workforce, orders were given that the Erith factory be enlarged to 1,000,000ft^3 on 19 July 1915. By the end of October it was completed. A smaller expansion was also scheduled for Crayford, and was completed in record time. Such was the demand that the Erith plant was expanded again in June 1916, by having a second production line added, and by the end of the year this was actually turning out guns faster than the original line had been.

Although the shortage of workers was gradually reduced with the increasing use of female labour (of its 107,000 employees, by late 1917 32,500 were female), it was never entirely solved. In a move many years ahead of its time, Vickers built housing for its workers, and provided day-care for their children, as well as a heavily subsidised canteen, theatres, sports fields and staff medical and welfare departments. By early 1916 both sites were in full production, and the guns began to roll off the production lines. Figures for the combined Vickers output of the two factories makes interesting reading as new employees became more proficient:

   1914: 339
   1915: 2,433
   1916: 7,468
   1917: 21,751
   1918: 41,699.

Erith guns were prefixed 'L' and Crayford were 'C', but also D, F and G. Specific factory production numbers and dates can be determined from the lettering and numbering sequences of the guns, which is complex, but a full explanation can be found in Dolf Goldsmith's seminal work, *The Grand Old Lady of No Man's Land*.

The physical manufacture of the guns was both labour intensive and slower than it could have been, as every part was scrupulously checked at each machining stage. Parts were measured for accuracy of fit using one of a vast range of parts gauges – there was one for every component in the gun. These were manufactured by the most skilled toolmakers to the precise dimensions of the blueprint; if the part was oversized it could be remachined, if undersized, it was scrapped and that was wasted time and money. Vickers began a production-line technique first pioneered in the United States, whereby one operator would be solely responsible for the machining of a specific part. This ensured that they became both quick and accurate, and faster production could, of course, result in a wages bonus for exceeding the daily target. However, this meant that there was always a fine line between quality of workmanship and precision. Rejected items did not count towards a bonus, so accuracy was vital. Bearing in mind that the machining required to make a lock body from a solid block of steel required 36 separate operations and approximately 3 hours of machine time; rejection due to an error could be hugely costly.

As production speeded up, so the price of a completed Vickers gun fell, from £165 in 1914, to £125 in 1915, £100 in 1916 and £74 by the end of 1918. In fact, profit margins were not high, at roughly 11% per gun. As a production comparison, a Vickers gun cost the equivalent of 13 No 1 Mk III Lee-Enfield rifles, whereas a Lewis gun required the equivalent production cost and time to make nine rifles.

To further complicate matters, hard combat reports brought attention to a number of functional shortcomings. The standard Mk I barrel (and it is worth noting that Maxim and Vickers barrels were slightly different, and were not interchangeable) was rifled with five grooves, and a 1:10 left-hand twist. To conserve weight, the original barrels and the locating trunnions were machined out, a process that saved a few ounces in weight and was in practice unnecessary. The walls of the Mk I barrels proved not to be thick enough, so they expanded when hot, resulting in friction at the muzzle, where oiled asbestos string was used as a water seal between the barrel and water jacket orifice. This could *in extremis*, slow

LEFT The difference between the lighter Vickers lock (left) and the Maxim lock as used on the MG08. Dismantling the Vickers required one split locking pin to be removed whereas the Maxim lock required the removal of three, all highlighted by the arrows. The difference in size is obvious, the greater weight of the Maxim lock contributing to the slower rate of fire. *(Author)*

ABOVE A comparative view of a 1915-production MG08 (top) and 1944-manufactured Lithgow Vickers. The slimmer proportions of the Vickers receiver are clear. The single long top cover on the MG08 is open and the two vertical trunnions for securing it in its sledge mount are obvious. The Vickers double top-cover system made access to the lock and feed far easier. *(Author)*

LEFT Reversing the crosshead on the Mk IV tripod enabled a reasonably high angle of fire to be achieved. The initial lack of a proper AA mount was a serious drawback at the front, making it difficult to engage the many enemy aircraft. Firing at an acute angle often resulted in feed stoppages as the belt twisted. *(Author)*

the rate of fire, but more seriously it caused the slide-on pattern of muzzle cup to jam, requiring brute strength and a mallet to remove. So, in late 1915, an improved barrel, the Mk II, was introduced which was thicker walled and had a threaded muzzle for a new type of muzzle cup. These barrels were no longer machined down to save weight and though they were about 2lb heavier, they lasted longer. As both types of barrel were in service simultaneously, two types of muzzle cup had to be manufactured, although as the earlier barrels were used up, the Mk II had become the standard by late 1917.

A more serious issue was with the right sideplate, lightened by the Vickers engineers, which held the crank handle and check lever – the curved bracket that stopped the forward motion of the crank handle as the mechanism cycled. The sideplate was simply too thin and warped under heavy use, so from summer 1915 a thicker pattern was introduced, first at Erith then Crayford. This also had a sturdier check lever and they were gradually retrofitted to all service guns. An elevation stop bracket was added to the lower left sideplate to prevent it jamming on the tripod crosspiece if the gun was fired at high elevation. Two modifications that often cause dissent among Vickers enthusiasts was the introduction in 1918 of a cone-shaped armoured steel muzzle booster to replace the flat one. This was not due to a need for any improvement in its function but to prevent the front of the water jacket from being pierced by bullets or shell splinters, the muzzle flash being a popular target for enemy riflemen. The angle of the heavy cone ensured projectiles would glance off. (The Germans' MG08s faced the same problem, but they opted for an armoured disc that attached behind the muzzle booster.) At the same time, Vickers began to produce guns with a smooth pattern spun-steel water jacket, which was simpler to make and easier to patch if punctured. The downside was that it had to be at least 0.10in thick, otherwise it would dent badly, which added to the weight of the gun. It is often stated that these smooth-jacketed guns were not produced until post-war, but Vickers certainly manufactured a number prior to 1918 and many of these guns were in service by the end of the conflict, thus they do not purely signify post-First World War

LEFT The Mk III spare parts box, which accompanied the gun everywhere. *(Richard Fisher)*

manufacture. These changes had the effect of increasing the overall weight of a gun to 33lb, which was still some 7lb lighter than the MG08.

Vickers soon began to realise that it could not cope with demand by manufacturing all components in house, so from 1915 many spares were subcontracted. This was an entirely new venture for Vickers, who had been accustomed to controlling their entire manufacturing output, but this simply was not possible with the number of parts required for a gun, let alone the myriad accessories and spare parts. A Vickers gun alone comprised 231 components, from tiny screws to the barrel casing. The Mk IV tripod added another 101 and the spares box a further 60. Then there were the boxes for spare parts and tools, the transport chests, oil canisters, cleaning rods and sighting devices such as the Barr and Stroud rangefinder and the clinometers. In addition,

BELOW The internals of a spares box, showing (from left), two spare feed blocks, small parts tins, muzzle cup and discs, screwdriver, belt easing tool, cleaning flannelette, hammer and chamber gauge. *(Ian Durrant)*

ABOVE The contents of one small parts tin: new corks for plugging the water jacket, split pins, brass strips and rivets for belts, gauze cleaning patches, assorted springs and locking pins for the lock and trigger mechanism. *(Ian Durrant)*

ABOVE A second spares tin holds a muzzle plug, tapered water jacket plug, muzzle securing pin, left and right lock levers, fuse, adjuster and lock parts. The gunners would cram as many spares as possible into the tins, just in case. *(Ian Durrant)*

there were a myriad ammunition requirements, such as boxes, belts, spares and the loading machine. No single manufacturer could produce the amount of parts needed, so Vickers had to rely on delivery of parts on a 'just-in-time' basis, now of course, a commonplace practice in industry where stock inventories are virtually zero. At the time this was regarded as a radical method of ensuring smooth production, as long as it worked efficiently. So while the major gun components were manufactured within Vickers' factories, everything else was subcontracted. Some idea of the complexity and scale of supply can be gained by looking at a few of the items that were prone to breakage, rapid wear or were regarded, at least by the soldiers, as disposable.

Spring manufacture was highly specialised and a technically precise process, but springs were required on a near biblical scale. There were insufficient companies in Britain with the expertise to manufacture the numbers needed, so at the end of 1915 Vickers placed orders with Colt for 25,000, who did actually have the production capability to manufacture them. Although the muzzle cups were supposed to be removed and cleaned at regular intervals, it was hard work to chip off the baked carbon. As the gunners did not have to pay for the replacements, they were simply thrown away and new ones used. Nine companies were contracted to make 92,000 muzzle cups just between December 1915 and March 1917 and in total over half a million were manufactured during the war. Waring and Gillow, then as now a well-established furniture manufacturer, produced 50,000 woven ammunition belts between June and September 1916 alone, and Myers and Son of Birmingham produced a staggering 7 million steel ammunition links solely for air service guns between August 1916 and January 1917. This was the tip of the iceberg, for there were dozens of other accessories, such as the tools – hammers, pliers, screwdrivers, cleaning rods and so on – that had to be sourced from companies that were either already producing them commercially or could do so, and at least 52 subcontractors were employed to fulfil the spare parts orders.

Unlike the Germans, who issued a low-power ZF12 optical sight with each MG08, no standard-issue Vickers sights were ever produced except for air service. However, optical items, such as clinometers, rangefinders and binoculars, were manufactured by a small number of specialist firms such as Aldis, Atchison, Broadhurst, Beck, Ross and Zeiss among others. For the most part these items were released to machine gun units from stocks already held for issue to the Royal Artillery. It wasn't until the very end of the war that demand overcame supply and optical instrument contracts were specially issued for the Machine Gun Corps.

The production of guns and spares was one thing, but without tripods, a Vickers gun was virtually inoperable and there had to be a careful balance maintained between the output of both. In fact, the Ministry of Munitions had flagged up the manufacture of the Mk IV tripod as 'most critical' and in spring 1915 placed contracts with 12 suppliers for 6,987 complete tripods. At the beginning of the year not one of those companies contracted had even prepared tooling, but to their credit by the end of July 4,582 had been completed at a cost of £30 each. The tripods were the subject of much criticism during the war, as fighting conditions often made the heavy Mk IV impractical to use. If the man carrying the tripod was hit, or the tripod itself damaged, then the gun on its own could not be brought into action, so in 1915 a lightweight 'emergency' tripod, designed by Charles Sangster was patented. It was simple and loosely based upon the bipod fitted to the Lewis light machine gun; the Vickers variant had a sprung steel band with a screw-clamp that fitted over the barrel jacket just forward of the receiver, roughly at the central point of balance on the gun. The clamp had a cast mount underneath with three short legs. With its low profile and sturdy construction, it was strong enough to cope with bursts of automatic fire, but care had to be taken that it did not move backwards under recoil, as this movement hampered the short-recoil action of the gun, and led to No 1 stoppages, as the fired case would not eject properly. The War Department made it clear that these were a battle expediency and '. . . not intended to replace the Mk IV tripod and will in future be issued on a scale of 1 per machine gun to cavalry and infantry units'. Six companies manufactured the Sangster, but despite the numbers produced (19,500), today they are one of the rarest of all surviving Vickers accessories.

Other tripod designs appeared, in particular ones specifically produced to cope with the needs of trench warfare, where gunners could not expose their heads without receiving attention from a sniper. Several devices appeared that enabled a Vickers to be fired remotely from beneath the parapet by a gunner, but the most effective was the Youlten Hyposcope Mk I, a variation of the simple box periscopes found in huge numbers in the trenches. A pattern mounted on the Enfield rifle had been employed for sniping at Gallipoli, but it was not accurate over about 50yd. A variant for the Vickers had a box-enclosed top and bottom mirror to provide a reflected view of the sights, and this clamped to the handles of the gun. A wire rod attached to the trigger bar ran downwards to the bottom of the box, where

ABOVE Two clinometers, Mk I and Mk II with a spirit level, all 1916 dated. With the eventual introduction of the dial sight, these were mostly phased out. *(Richard Fisher)*

LEFT A remarkably accurate drawing of a Vickers, with canvas condenser bag and Sangster mount, is used to advertise the MGC Comforts Fund. *(Lawrence Brown)*

RIGHT A demonstration of an early Youlten Hyposcope, a simple box periscope strapped to the firing handles of the Vickers. As accuracy was not a prerequisite, it worked tolerably well, but was a tempting target for German snipers. *(Author)*

the firer's hand could depress another bar and fire the gun. It was crude, but efficient, and on a machine gun accuracy was of no consequence. A more sophisticated design soon followed, with a prism aiming device attached to two metal tubes that slid up over the firing handles and dropped down roughly level with the bottom of the crosspiece. A centre tube held a firing trigger which attached to the firing mechanism on the rear of the gun. The gunner looked through another prism sight by his left hand to aim. By the end of 1916 they were declared obsolete and very few exist today.

As the war progressed and the nature of the fighting changed, it became clear that the increasing use by machine gun crews of concrete emplacements or 'pillboxes' had created its own problems. The issue tripod was often difficult to set up inside the confined space of an emplacement, but worse was the limitation imposed by having the traversing centre of the gun underneath the receiver. In practice this meant that it was extremely hard to get a reasonable arc of fire from within the narrow loopholes of a shelter, as the muzzle was set too far back and visibility was restricted by the edges of the aperture.

The 'Mounting, Trench, MG, Mk 1' was an attempt to rectify this and was the result of a design by a machine gun sergeant called Longfield. It comprised a short three-legged framework whose two forward legs sat flat against the wall of a bunker beneath the embrasure, with the gun sitting on top of a flat C-shaped cradle. At the front arm of the cradle was a free-rotating pivot to which the gun was attached, while the rear mount ran along a short rail on a machined slider that meant the gunner could move the gun laterally, providing a 77° arc of fire, as the traversing point was now virtually under the muzzle. To the right was

BELOW A Mk II Hyposcope at Gallipoli, far more sophisticated than the early pattern. It has a mirror clamped to the top of the receiver and twin metal supports attached to a trigger, providing a far steadier firing platform. *(Author)*

a metal frame to hold the ammunition box. It worked well despite its 100lb weight, but there was an unexpected side effect to using the Vickers in confined spaces – something that had become apparent to German machine gunners quite early on. When fired from an enclosed position, the muzzle fumes would be sucked back into the pillbox. Over a period of time these could make the crew physically sick (as happened in tanks) so various methods of prevention were undertaken.

Flash hiders were tried, these being no more than lengths of drainpipe attached to the muzzle, developed for trench use but without great success. In part this was because a build-up of gas would cause an almighty explosion to erupt every few dozen rounds, giving away the position of the gun and, in the dark, blinding the crew. Although they did deflect some of the gas discharge, it was felt that a purpose-made attachment was required.

In late 1918 the experimental 'Deflector, Blast, Vickers .303 inch MG, Mk 1' was created. It was a steel cup about the size of half a tennis ball, with a tube attached that slid over the muzzle booster of the gun, and it worked – after a fashion. It did deflect the gas forwards but did nothing to conceal muzzle flash. Unlike many of the First World War's other contrivances, this one was not to vanish without trace and was to be resurrected, as we shall see.

There were other vital but frequently unconsidered pieces of equipment produced for the guns, many of which were connected in one way or another with the ammunition, its feed and the requirements for reloading. There were small things like the belt-expanding tool that eased the pockets on belts to enable cartridges to be correctly inserted; vitally important if belts had got wet or were brand new and the pockets very tight. There were also belt-filling machines, which resembled an old-fashioned meat-mincer, that were needed in large numbers but were both complicated

RIGHT **A belt-loading machine. The hopper for loose cartridges is on top, the crank handle for inserting the ammunition into the belt at the lower left. The solidity of its cast steel base can be clearly seen.** *(Richard Fisher)*

to make and slow to manufacture. Each one weighed almost 23lb and required a hopper to be filled with loose cartridges that dropped down a chute into a charger guide operated by a pawl, which then inserted them correctly into the belt. As it needed careful setting up on a firm, flat surface it sometimes proved to be less than practical on the battlefield, so a portable model was trialled that could be held on the knee, and though much cheaper to manufacture, it was never adopted. Anyone who has ever attempted the painful process of loading a Vickers belt by hand will understand how important these machines were.

ABOVE **An extremely rare Mk III aiming lamp, first introduced in October 1919, to provide direction for night-firing, visible only from the gun position. It was to remain in service until 1968.** *(Richard Fisher)*

RIGHT The 'Mounting, Tripod, Auxiliary, MG, Mk III', otherwise known as the Sangster. The leather securing strap sits at the front of the jacket, but this early pattern has pointed feet, which sank easily in soft ground. Later ones had flat pads added. *(Richard Fisher)*

BELOW A loaded belt; the three cartridges on the left are correctly seated, the fourth is too low and the fifth and sixth too high. The latter three would all cause feed stoppages. *(Author)*

Neither should the belts themselves be overlooked. The standard 250-round cloth belt initially developed by Maxim was specifically designed to fulfil the function of enabling ammunition to be reliably loaded into the gun. First World War belts were all distinguished by their long brass feed tabs, normally bearing the maker's name and three-eyelet brass spacers, with every third spacer having a forward projection. There were several very specific reasons for this design.

Firstly, the spacers gave the belt rigidity, a belt being fired will flap about wildly as it comes out of the box, and if the fabric is too pliable it will literally tie itself in a knot. Secondly, the projections showed the exact length that a cartridge must be seated in the belt; to feed properly into the gun, the bullet tip must be level with the tip of the spacer. Lastly, they had a lesser-known function of preventing the cartridges from slipping out of their belts if they were subjected to rough handling during transit. Opening a new box in the heat of battle to find 30 or 40 rounds lying loose inside was not something that any machine gunner wanted to deal with.

Then there was the problem of selecting the right fabric for belts. Being cotton-based it absorbed moisture very quickly, doubling the weight of the loaded belt, placing a great strain on the feedblock pawl and shrinking the cartridge loops so that the carrier sometimes failed to extract a fresh round from the belt to load it, but all other materials lacked the combination of stiffness and malleability required. It proved impossible to satisfactorily waterproof them. Paraffin wax helped to a degree, but when hot it collected dirt and grit so attempts were made to find an alternative solution, and something less gas absorbent seemed logical. Metal seemed the most obvious material, and Thomas B. Sangster (no relation to the Sangster of tripod fame) designed a system of non-disintegrating metal links that were attached by split-pins to each other. Although it was a clever concept, they proved difficult and expensive to make, failed to grip the cartridges properly as the belt fed into the gun and also allowed the rounds to work loose in transport. These, however, must not be confused with the air service links, which were of a totally different type and will be examined in the next chapter. It was, however, the first step in the right direction, for today disintegrating metal links are universally used on machine guns. Various other forms of cloth belts were tried – some with spacers, some without and with no projecting eyelets – but nothing seemed to work as well as the original design.

One of the imperatives for ensuring reliable performance with the Vickers was in ensuring the belts were kept dry. This was a fine theory, but nearly impossible to achieve in practice using standard ammunition boxes. The first boxes for Maxim guns were of teak, beautifully made and, containing 250 rounds of .577/450-calibre ammunition, extremely heavy. With the introduction of the .303in service round

in 1890 a smaller box was introduced and the belt dimensions correspondingly reduced.

A 'Box, Belt, Ammunition, Machine Gun, No 3, Mk I' was approved in February 1897 and this was constructed of pine, the full-length hinged lid opening at one end and secured by a recessed metal catch with a leather carrying strap. A Mk II variant followed that had metal grooves inlet into the ends to allow the box to be placed in a mounting bracket. The problem was that none of these were waterproof and, of course, no thought had been given to a weapon that had yet to be invented – gas.

Exposure to the chlorine gas that was introduced from April 1915 resulted in verdigris forming on the ammunition and when fired from a hot barrel, it turned into a form of cement, eventually jamming a fired cartridge so solidly that it would fail to extract and would lock the breech solid, or result in a separated case. In order to try to solve this, the 'Box, No 3 Mk III' was introduced and this was to become one of the most commonplace boxes during the First World War. It differed in having a tinned metal lining that held the cartridges more securely in the belt and a leather handle was riveted to each side of the box. Service in the flooded trenches of Flanders soon showed up the deficiencies in their design and construction, however, for the liners did little to stop gas ingress or reduce water penetration once the lid was open. Indeed, so bad were the problems caused by wet belts that rainguards were issued to try to stop opened boxes getting wet, while hapless gunners were advised to avoid exposing their boxes or belts to water, an impossibility most of the time. Oiled flannelette patches of the type used to clean the barrels were placed under the lids in an attempt to reduce the effect of both gas and water. The answer was simply to make boxes out of a waterproof material, so after much experimentation at Enfield, an entirely new design of metal box was introduced in October 1915.

This was the 'Box, Belt, Ammunition MG No 6 Mk I'. It was made of welded galvanised steel plate with a double-hinged lid that had felt strips glued to the underside to stop gas penetration. The top lid still hinged at one end, but also had a secondary centre hinge

ABOVE Five different patterns of the No 3 ammunition boxes, from the early one-piece lid to the late war double-hinged pattern. *(Richard Fisher)*

that enabled the lid to be half-opened for the belt to be extracted, but without unnecessary exposure to the weather. It too had a double leather handle. From early 1916 the new boxes were to be issued according to the Army's List of Changes, No. 17669. '. . . In the proportion of four steel to ten wooden boxes per gun, and officers of units having Maxim or Vickers guns on their charge will submit indents reporting the numbers of wooden boxes currently held accordingly, returning all surplus wood boxes to store.' It is doubtful if many of the old boxes ever made it back to the stores, for they provided an excellent source of firewood for the gunners, which is no doubt why examples are so scarce today. Inevitably improvements were made: strengthening the body to prevent crushing when stacked (a filled box weighed 21lb 9oz), smoothing the exterior surface, using webbing for the handles and completely lining the lid with thick felt for the prevention of gas ingress.

New patterns came thick and fast; No 7, No 8 and No 9 Mk I boxes all came hot on each other's heels, although work on improving them continued after hostilities ended. Despite remaining in service until at least 1949, equally rare are the 'Belt Box Carriers, .303 inch', metal frames designed to carry an ammunition box and

LEFT A No 7 Mk I steel box with the felt packing in the lid to try to prevent gas and water ingress. The lid is half-opened. *(Author)*

BELOW The improved No 8 Mk I box showing the felt inner lining, adopted in November 1917. Filled, it weighed a little short of 22lb. *(Richard Fisher)*

keep it off the ground. These simple affairs were no more than a welded steel basket that hooked on to the right side of the Mk IV tripod crosshead to improve ammunition feed and aid traversing.

Vickers guns were, of course, water-cooled – the barrel holding almost 7½pt, which weighed 10lb – and the cooling was by means of a constant-loss system. After 600 rounds of rapid fire, the barrel heated up to the point where the water temperature reached 100°C; the water then turned to steam and this required a method of venting. To achieve this, a hose was attached to the outlet at the lower front left of the barrel jacket. On First World War guns this was screw-threaded, the hose being a 6ft-long, flexible armoured type with a wing nut for attachment. The threaded boss was very vulnerable to damage that could prevent the hose from being attached, effectively rendering the gun useless. Neither did the armoured hoses prove entirely satisfactory as they became completely frozen solid in cold weather and turned rusty in wet weather – as well as being difficult to handle and stow. This was not rectified until virtually the end of the war as until then no suitable rubber had been produced that could stand the boiling temperatures and 40psi pressure that was generated. Eventually a reinforced black rubber hose was issued. The steam in the jacket clearly had to be directed somewhere, so a standard 2-gallon motor petrol can was issued. These were soldered from thin steel sheeting, with a carry handle on top and large brass screw filler cap. They were somewhat fragile but cheap and plentiful. To aid filling the gun via the small aperture on the top right of the jacket, a variety of funnels were included with the Vickers parts kit: collapsible leather or canvas, solid tin, as well as several patterns of spout that screwed on to the filler neck of the can. The handiest was the latter, which were often attached by a short chain to the handle, so they could not be lost in

RIGHT A close-up of the modified steam hose connector, which was less susceptible to damage and more positive to use than the screw pattern. The armoured nose cone was a standard feature on all Second World War guns. *(Ian Durrant)*

BELOW A soldered patch on the jacket of a gun. In fact, this particular Vickers has so many patches it is marked 'EY' for emergency use, and was primarily employed for drill purposes. *(Richard Fisher)*

RIGHT A Second World War jacket repair kit comprising a malleable patch, metal cover plate and adjustable steel clamping bands. *(Ian Durrant)*

LEFT An Indian Army Vickers gunner poses with a .303 Maxim somewhere on the Western Front. The gun is an early M1893 rifle-calibre Maxim and wears a leather cover on the jacket. Although not officially issued at this time, many were produced by the farriers of the unit. *(Author)*

transit. The cans were normally half filled with cold water to prevent too much steam from escaping as it condensed and also to provide an instant source of water for replenishment. It is also worth mentioning one of the greatest problems that beset all these water-cooled guns, which was the vulnerability of the jacket to puncturing from shrapnel, shell splinters or bullets. Small holes, such as those made by bullets, for example, could be plugged with plasticine and wrapped with cleaning flannelette. In the 1930s, a repair kit was supplied and this comprised malleable metal patches with rubber sealing pads that were held in place by soft metal straps, tightened by thumb screws. Guns can sometimes be seen in museums with soldered patches, which were the armourer's preferred method of repair, but this was impossible to be applied to a gun that was damaged while it was in the line. Indeed, one of the reasons for the decision to make smooth jackets in 1918 was the difficulty in repairing fluted ones. If a jacket was too badly holed to repair, it was possible to continue firing in short bursts, but the barrel had to be allowed to cool. Machine gunners stated that 500 rounds of rapid fire would heat an uncooled barrel to red hot, at which point the gun would simply cease to function due to the reduced gas pressure as a result of the rifling being worn out.

Heat was a constant problem for the gunners, as changing a barrel meant handling a boiling-hot jacket and a near-red-hot barrel. Extremely nasty burns could be inflicted through touching virtually any part of the gun without wearing protection. 'Gloves, Machine Gunner' were made of asbestos wool with the leather pad on the palm. A slit in the palm enabled a single finger to protrude that could operate the safety catch, though many gunners carried more practical leather gloves in their battledress pockets.

In complete contrast to the heat was the problem generated by extreme cold, which would quickly freeze the water in the jacket and coagulate oil on working parts, making them at best sluggish to function and at worst locking the gun up totally. Jackets were either wrapped in blankets or sandbag covers packed with straw if it was available. If the gun was in a place where enemy action was expected, the teams would fire about 20 rounds at regular intervals, to keep the water warm. The manual stated that up to 33% glycerine could be added to the water (assuming it was available) to prevent freezing with a maximum of 5pt of coolant retained in the jacket. If this was not possible, it was suggested that the men sleep with the gun between them, one keeping a lock in his pocket to ensure it was functioning. On extremely cold nights the jacket would be drained of water on the basis that, should the need arise, a belt or two could be fired immediately and the water topped up in the interim.

Cleaning the guns was a straightforward process. The barrel could be left in place or removed, and a brass D-handled cleaning rod

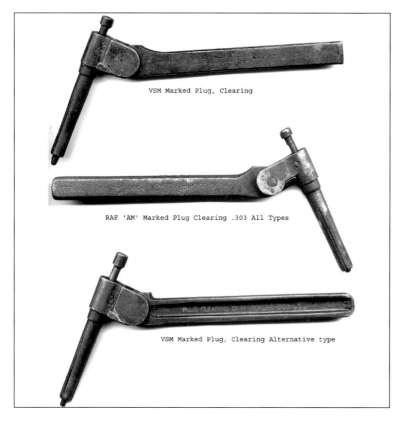

ABOVE Three patterns of .303in clearing plugs. The pointed end was inserted into the broken case in the chamber and the projecting pin tapped tightly to expand the plug body, theoretically enabling the case to be levered out. *(H. Woodend)*

BELOW Leather spares wallet containing combination tool, clearing plug, oil can, spring balance, lock, fusee spring, pliers and anything else the gunner thought was handy. *(Richard Fisher)*

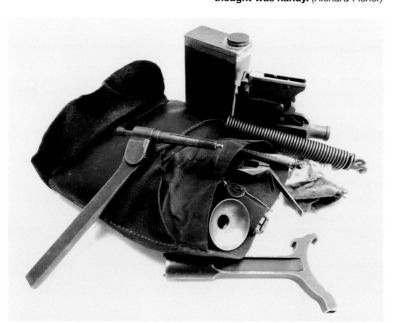

'Rod, Cleaning .303in MG Mk I' was supplied, along with flannelette patches usually referred to as 'four by two' (from its size in inches), although the manual helpfully recommended small strips of Army blanket as being 'very effective'. Gauze patches were used for removing heavy fouling from barrels, and solvents such as petrol or paraffin were useful if available. For thorough cleaning with the barrel removed and the muzzle plugged, boiling water was poured into it, the heat loosening up the fouling, prior to gauze patches being run through. Cleaning could be undertaken from either the breech or muzzle end of the barrel as, unlike rifles, wear to the muzzle crown that affected their accuracy was not an issue with a machine gun, as the barrels were frequently replaced. The barrels were gauged for throat erosion caused by heat and a go/no go tool was supplied to check this. Badly worn chambers meant the barrel had to be scrapped, even if the rifling appeared to be in a good state. The muzzle cups might be cleaned if no others were available, but normally a new one was fitted. Oil was carried in a number of different patterns of small tins,

BELOW A nice photo of a gun team cleaning their Vickers at Zillebeke, near Ypres, in early 1915. The tube lying on the parapet at top left is an early 'drainpipe', an attempt to reduce the muzzle flash from the gun. Just visible behind it is a box periscope, possibly a Youlten. *(Author)*

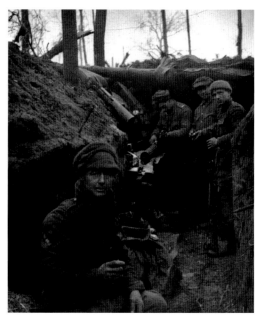

LEFT A Barr and Stroud No 12 rangefinder on its tripod. One was issued to each four-gun section and men were specially trained as 'rangefinders'. Although best suited to calculating range for direct fire, it could also be used for indirect fire. *(Richard Fisher)*

as well as in the hollow spade-grips of the gun, both for lubricating internal components and cleaning barrels. All components were removed and washed in solvent then dried and oiled. When out of action, a gun would be totally stripped, and remarkably few tools were required to achieve this. A long-handled C-shaped combination tool was provided but the design of the guns required little else aside from a small hammer and punch to strip the lock. A small number of special tools were issued, the most common being the 'Plug, Clearing, Vickers .303 inch' or more simply 'the case extractor', to remove stuck cases from the chamber. When not in service, the gun was packed into a wooden carrying case, with a spare barrel and cleaning rod, which had rope handles at each end – necessary in view of its 38lb weight.

Hitting your target was a fairly vital requirement, and considerable effort was put into achieving what appears to be an apparently straightforward task. The most widely found item was the clinometer, a combined level gauge and angle-of-sight tool. Most were manufactured pre-1917 by J.H. Steward of London, who had been supplying the Royal Artillery for some 20 years. Using one was a simple process. The instrument was placed on a nearby surface or held steady and levelled using the spirit bubble, then it would be 'aimed' at the target (or an artificial aiming mark) and a reading taken. It would subsequently be placed on top of the gun's receiver and the gun levelled to match the reading in the spirit bubble. After September 1917, adapted clinometers with special rails fitted on to the receiver sideplates were produced, modified from existing artillery clinometers, and used with separate 'directors'. The next most commonplace instrument was the infantry rangefinder manufactured by Barr and Stroud, but unlike the simple clinometer, using it was a specialist task. It provided a split image similar to that still found today on some cameras and required four readings to be taken for an average to be calculated (the level of accuracy required was for machine guns, not the precision of a sniper). The preferable target for this was something vertical, such as a tree or telegraph pole, as the stadia in the rangefinder were two vertical pointers. It was 32in long and was normally mounted on a sturdy tripod, but in action it was a matter of selecting suitably visible objects like a copse, haystack or building and marking them on a range card along with their distance. This was always handed over to a replacement gun crew so time did not have to be wasted in duplication of effort. Range tables were also supplied, and this will be covered in the section related to ammunition in Chapter 5.

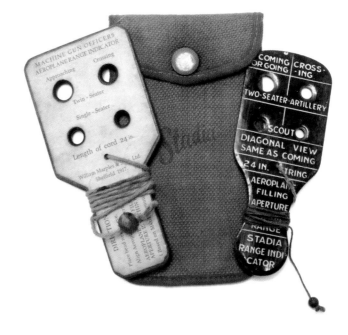

BELOW Two examples of the stadia target indicators for anti-aircraft ranging. If the plane filled the aperture, it was within range. Home-made ones were often used, based on a template supplied by the Small Arms School. *(Richard Fisher)*

67

FIRST WORLD WAR MANUFACTURE

Chapter Five

Second World War manufacture and improvements

By 1937, of the 58,000 Vickers guns manufactured in the First World War almost 30,000 had been disposed of. The remainder had to be quickly reconditioned and production resumed. Few military commanders expected the old guns to play as big a role during the forthcoming war as in the last – but they were to be proved wrong.

OPPOSITE A Gurkha gunner uses the issue cleaning rod to swab out a barrel during the Italian campaign, late 1944. The Vickers appears to be a refurbished First World War model. *(IWM NA22724)*

When the ceasefire began on 11 November 1918, some 56,057 Vickers guns had been manufactured: 37,239 at Erith and 18,818 at Crayford. The vast majority of these were still in the hands of Machine Gun Corps (MGC) units across the battlefields of Europe. They had operated reliably throughout the war and been in service with a dozen countries on land, sea and in the air. An incalculable amount of time and investment had gone into their design and construction and despite the disbanding of the MGC in 1922, there was no doubt in the mind of the British Army that the guns would continue to play a vital part in any future wars. Of the Vickers guns left on inventory post-war some 30,000 were sold off or returned home with Dominion troops, but the number of working guns left in store or in the hands of infantry regiments was still considered to be sufficient to equip the Army.

So, while the manufacture of weapons of all types was dramatically scaled down, work continued on improving the Vickers and its ancillary components. This was primarily undertaken at Enfield as, in the 1920s, Vickers had begun to expand into tank and aircraft production, and in 1927 they had merged with the Armstrong Whitworth Company. Indeed, by 1939 a quarter of all the tanks in use by the British Army were manufactured by them and they had also moved into large-scale aircraft production, acquiring the Supermarine plant that manufactured Spitfires. However, there was still a requirement for machine guns and parts, but by then only Vickers still had the tooling and plant, Enfield having ceased gun production. Throughout the 1920s this was of low importance in terms of equipping Britain's relatively small peacetime Army, but in the early 1930s the threat of another war with Germany began to become a reality. This time, though, the British Government was not going to be caught napping, and plans were put in place to resume the manufacture of Vickers guns, as there were now considered to be insufficient numbers to equip a larger Army.

Consequently, in early 1937 a contract was placed with Crayford for 12,000 new guns as well as 13,207 locks and 64,138 barrels, Erith having been prematurely closed at the end of 1931, while the plant at Crayford also produced inter-war commercial guns for international sales. Sideplates and barrels were partly fabricated at a new 'shadow' production site in Bath, which was considered to be safe from any possible enemy bombing, and all were of the modified Mk II pattern. These guns can be recognised by their 'V' prefix, which was given to guns up to number 9,999; thereafter a 'W' was used. Almost all major components were manufactured by Vickers or the by now re-established RSAF Enfield. Between them, they produced most of the 19,205 Mk IV tripods that were also made, but subcontractors, who had produced tripod parts during the period 1914–18, were also employed. These were identical to earlier patterns, although the dial plates were sometimes smaller, and the crosspieces could be made of either alloy steel or brass. For small specialist components such as screws, pins, springs and brackets, some 27 subcontractors were employed, and an unrecorded number supplied the toolkits.

In terms of mechanical function, these guns were identical to those of 1914–18. The only visible difference was that they were now manufactured with a smooth water jacket. Confusingly for Vickers historians, some of the first production run were made using up old stocks of corrugated metal, their appearance belying the fact they were actually new production guns. Despite the proven effectiveness of the guns in combat, there appeared to be little desire to leave things alone on the part of the engineers at Vickers and RSAF Enfield, and a myriad changes were made to components – mostly, it must be emphasised, to simplify production. The list is lengthy, but the most important changes encompassed modification to the gunmetal feedblocks, now manufactured with some steel parts to cut down on wear, and access was made easier to reach the belt-holding pawls, the firing pin and lock spring. The carrier was simplified, the firing handles and trigger assemblies improved and the trunnion block lightened. The tangent rear sight went through several modifications, eventually becoming the 'Sight, Tangent, Mk IV', with the large thumbwheel adjuster being replaced by a sprung-loaded type with the 'V' notch of the rear supplanted by an aperture

RIGHT A very atmospheric photograph of a well-dug-in Vickers of the 8th Battalion, Middlesex Regiment, on the Siegfried Line, 8 February 1945. They are using earth-filled ammunition cases to provide extra protection for their dugout. The amount of ammunition expended can be gauged from the empty belts piled up. *(IWM BU1785)*

pattern, set within a protective shield. All guns were produced with the cone type of muzzle deflector, although many are seen during the Second World War with the old flat-nosed pattern, presumably because it was still available in quantity.

To address the constant complaint about the heat of the water jackets, in about 1935 a heavy canvas cover was introduced, which was laced underneath to hold it in place. It became an almost permanent addition to Second World War guns. The other most commonly fitted accessory from 1940 onwards was the new 'Eliminator, Flash and Blast Deflector, Vickers .303 inch MG, Mk I', generally referred to as the 'parabolic' deflector. This was created by the simple expedient of soldering two of the old pattern together, the main difference being that the new, rather ungainly item actually worked very well. Virtually all Vickers guns photographed in service from this date onwards have the flash eliminator fitted. There were many other improvements to accessories as well, so many, in fact, that covering all of them in detail here would be impossible.

Some of the more useful included a lightweight magnesium Mk 5 tripod that was introduced towards the end of 1945, as well as a bipod and monopod, to enable vehicle guns to be used in a dismounted role. A twin-gun mount for an anti-aircraft role was made in 1942, and this was widely adopted by the

RIGHT Machine gunners of the 2nd Middlesex Regiment providing covering fire for an attack on Overloon, Holland, October 1944. The gun has been fixed rigidly in position to provide high-angle fire, and the gunner is merely there to hold the trigger down. Note the plain woven belts. *(IWM B010814)*

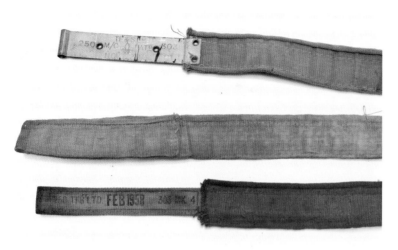

ABOVE Examples of post-First World War stripless ammunition belts. From the top: Mk III with brass tab, made in November 1939 (impregnated with a solution to make it rot-proof, hence the green colouring); a Mk IV belt with no tab; and a late Mk 4 (the Army changed to Arabic numerals in 1944) manufactured in February 1958. *(Richard Fisher)*

ABOVE RIGHT Exhausted Australian gunners take a breather in Borneo during the 1945 advance. The hose has been wrapped around the jacket, a favourite method of stowing it when the gun was out of use. The brass D-handle of the cleaning rod can just be seen tucked into the water jacket cover. *(Australian War Memorial 110981)*

BELOW A Mk 7 tin, showing the closely packed belt, with the brass spacers ensuring that the ammunition did not move in transit. Later stripless belts had cardboard packing inserted to do the same job. *(Author)*

Navy. Belts and boxes did not escape attention either – in fact, finding a cheap yet efficient belt seems to have been a constant obsession since the Vickers first went into service. The brass-stripped belts were now deemed too costly to make and while plain belts had been tried, they had never worked satisfactorily. Post-1918 these surviving 'Belts, Stripless, Vickers .303 inch Mk I' had been dumped in Army stores and forgotten about, but many were rescued for testing in the 1930s as renewed efforts were made to produce an acceptable type of plain belt. It was found that a thinner type of cotton webbing, manufactured in continuous lengths and heavily stitched, worked quite satisfactorily. Neither did they require the brass feed tabs, most utilising a thick webbing tab instead, on which the maker's details can be found. Each belt was cut into 251-round lengths before being filled. They were designated 'Belt, Vickers .303 inch Mk IV'. This plain pattern became the standard for the Second World War and they were still in use at the time the Vickers was taken out of service in the 1960s.

Other types, some still with brass feed tabs, were also supplied by American and Canadian contractors. If a solution had finally been found for the belts, finding the perfect ammunition box was still an ongoing task. Metal was by now the clear choice for protection of the ammunition from damp and gas, but it needed to be sealed to preserve the cartridges when in store, which could be several years. The solution eventually arrived at was a soldered can with a handle that ripped the lid open, much like a modern tin can, exposing the belt. Each of these weighed 24lb.

RIGHT This classic photo of a Vickers at El Alamein must have been specially posed for the camera, as there is no water hose or tin, although judging from the movement on the belt, the gun is apparently firing. *(Author)*

CENTRE Variants of Vickers belts, from the top: A standard Mk I belt, used until the Second World War, then a 1918 stripless cotton belt that had to be hand-filled. Next is a length of 'Sangster' non-disintegrating steel link and finally a Prideaux Mk III aircraft linkage. To the right is the early pattern belt tool – later ones were all wood. *(Richard Fisher)*

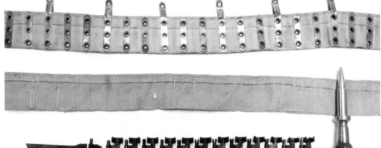

They were loaded and sealed at the factory and then transported in wooden outer boxes. As belts were now free of brass spacers and had more space around them, they were loaded into the inner box with nine thick pieces of cardboard packing to ensure they stayed in place during transit. Steel chests holding two boxes each were also produced, weighing a hefty 62lb each. When packed with earth, these provided excellent protection for dug-in gun crews and this was to become the standard system for all small-arms ammunition (SAA) until the 1970s.

Ammunition

While much is written about the history and development of the guns themselves, the most critical of factors, the ammunition, is often glossed over. From the outset, the Vickers was designed to fire the Mk VII .303in cartridge. This was a pointed bullet with a flat base that weighed 174 grains, and was propelled by 37.5 grains of cordite (which had excellent storage properties), providing it with a muzzle velocity of 2,440fps and maximum range of 2,900yd. Because factories could not keep up with demand for ammunition (6 billion rounds of .303 were manufactured between 1914 and 1918), from June 1916 a Mk VIIz cartridge was also manufactured in the USA and Canada, the 'z' signifying nitrocellulose loading. However, there was a tactical requirement for the range of the Vickers to be increased, particularly for indirect or plunging fire, but no attempts had been made to develop a new bullet. This

LEFT The four most commonly used Maxim machine-gun rounds in both world wars. From the left: British .303in Mk VII and Mk VIIIz; Russian 7.62 × 54mm R; and German 7.92 × 57mm Mauser. *(Author)*

ABOVE The brass feed-tab on a Vickers Sons and Maxim-made belt. *(Author)*

BELOW There were no automated processes for belt filling, which was done in the factory by teams of women 'fillers' using small machines, not visible in this picture. Any obviously defective cartridges would be thrown to one side. Bottom right are large ammunition boxes for Mk VII .303in cartridges. *(Author)*

RIGHT Bullet development. From the left: an 8mm French Balle D, the first pointed, streamlined bullet; British Mk VII .303in bullet, with parallel body and flat base; and British Mk VIIIz streamlined boat-tailed bullet. *(Author)*

is somewhat strange as all the other First World War protagonists were using boat-tailed bullets, which improved range and stability, and producing a similar pattern for the .303 would not have been difficult. Be that as it may, it was not until late 1939 that the Mk VIIIz was introduced, loaded with 36.5 grains of Neonite, a nitrocellulose derivative, raising the velocity by 100fps, and with its streamlined 175-grain bullet it increased the range of the Vickers to 3,700yd. One important factor discovered during testing was that the same barrels must not be fired with both Mk VII and Mk VIIIz ammunition, as the different burn temperatures rendered the barrels inaccurate very quickly due to premature bore wear. Neither should Mk VIIIz be issued to the infantry, as there was a slight danger of barrel damage if fired from a service rifle.

Some explanation of the power of the standard service ball round would not go amiss here, for most soldiers failed to understand its destructive power. A .303in bullet, with a velocity of 2,600fps (almost 1,800mph) produces almost 3,000ft/lb, or a little over a ton and a half of energy and can penetrate $\frac{1}{3}$in of steel plate, 8in of oak, two courses of mortared brick or 18in of hard-packed sandbags at 200yd. Bearing in mind that the human body is roughly equivalent in density to a bar of soap, it offered virtually no resistance to bullets, even when fired from long ranges. At the most optimum combat ranges, between 300 and 600yd, a single bullet was capable of penetrating the soft tissue of several individuals before losing sufficient velocity to cease being lethal. Frequently they would strike bone and tumble, turning into wildly spinning pieces of deformed metal and creating dreadful wounds.

For the infantry, two other types of cartridge were employed alongside the standard ball – tracer and armour-piercing. Finding a suitable tracing compound was not easy; it should not burn too fast or too hot and the Mk IL, a pattern introduced pre-1914, was found to be ineffectual when fired in combat, igniting too soon and burning out too quickly. After experiments at the factory of Aerators, who produced the gas canisters for recharging soda water bottles, it was discovered by mid-1916 that a compound comprising eight parts barium peroxide to one part magnesium was ideal. Although approved quickly by the Ministry in October 1916, it was initially only supplied for air service, because the bullets generated so much heat that barrels burned out too quickly on ground-employed Vickers. In June 1917, a new pattern, the SPG Mk VIIT was introduced. This ignited shortly after leaving the barrel and was deemed more suitable for infantry machine guns, so by late 1917 tracer ammunition was being supplied at the ratio of 250 rounds per 5,000 ball. Most gun crews used it sparingly, as it could give away the position of an emplacement, but when fired in a mixed belt of one-in-ten tracer to ball it was invaluable for ranging and also in an anti-aircraft role as it enabled gunners to correct their aim. They dryly referred to it as the 'Sparklets' cartridge – after the soda syphon of the same name. Further development by the Royal Laboratories at Woolwich continued through the 1930s and in 1939 a new cartridge, the 'Tracer, .303 inch G Mk II L' was introduced with a lighter 154-grain bullet, burning with a red trace. This progressed quickly from the Mk II of 1943 to the final incarnation of the tracer cartridge, the Mk 7 of mid-1945.

The second most commonly found type of special ammunition – armour-piercing – had initially been developed in the late 1880s by Helge Palmcrantz of Nordenfelt fame, but although the Navy was interested, the British Army saw no application for it. Their opinion changed, however, once Germany introduced the SmK AP round in late 1915. Britain rushed through a new cartridge, the W Mk VII AP, with a steel core inside its 174-grain bullet. It was almost identical to the ball cartridge, but could penetrate almost ½in of armour plate at 100yd. While ideal for aircraft, its practical application for the infantry was limited, but it did prove effective in dealing with the armoured shields on German field guns, MG08s and the sniper plates that proliferated the parapets of enemy trenches. Post-war, it was further developed into the .303in W Mk I series and from 1941, with minor variations, these became the standard for machine gun service, the final pattern being introduced in 1945.

Even if the function of the gun and ammunition had been more or less perfected by the outbreak of the Second World War, there still remained the matter of hitting the target. The tactical role of the Vickers gun

ABOVE This photo illustrates very well the limited arc of fire that a Vickers had inside a blockhouse because the crosshead mounting was set so far back. *(Author)*

BELOW A standard Range Table for First World War Mk VII ammunition. *(Author)*

SECTION XIV.—RANGE TABLE FOR MARK VII AMMUNITION.

Muzzle velocity with Mark VII ammunition ... 2440/s.
Weight of bullet ... ... 174 grains.
Weight of charge, cordite ... ... 38 ,,

Range.	Angle of Elevation.		Range.	Angle of Elevation.	
Yards.	Degrees.	Minutes.	Yards.	Degrees.	Minutes.
100	—	3	1,600	2	35
200	—	7	1,700	2	57
300	—	11.5	1,800	3	21
400	—	16.5	1,900	3	47.5
500	—	22	2,000	4	16.5
600	—	28	2,100	4	48
700	—	35	2,200	5	22.5
800	—	43	2,300	6	—
900	—	52	2,400	6	41.5
1,000	1	2	2,500	7	27
1,100	1	13.5	2,600	8	16.5
1,200	1	26.5	2,700	9	11
1,300	1	41	2,800	10	10.5
1,400	1	57	2,900	11	15
1,500	2	15			

When firing Mark VI ammunition from a gun sighted for Mark VII, up to 600 yards an addition of 200 yards should be made to the observed range. Above that distance 250 yards should be added. The results should in all cases be checked by observation.

MARK VI AMM^N

TRAJECTORY DISCS.

ILLUSTRATING THE CONE OF FIRE AS IT WILL APPEAR IN OVER HEAD FIRE, ETC.

Range	Diameter of Discs.		Height of Centre of Disc above ground, Muzzle of Gun is taken as being 20 inches above ground.		
	75% Cone	100% Cone	700 yards Trajectory.	800 yards Trajectory.	900 yards Trajectory.
Yards	Ft. Ins.	Ft. Ins.	Ft. Ins.	Ft. Ins.	Ft. Ins.
100	8½	2 0	5 5	6 5	8 8½
200	1 3¼	3 6	8 6	10 6	12 10
300	2 0	5 0	10 2	13 4	16 10
400	2 8½	6 6	10 7	14 8	19 6
500	3 6	8 0	9 4	14 2	20 2
600	4 0	10 0	5 11	12 1	19 0
700	4 6	12 0	—	7 6	15 6
800	5 6	14 0	—	—	9 1
1,000	6 8	16 0	—	—	—
1,500	10 0	24 0	—	—	—
2,000	13 4	32 0	—	—	—

DEPTH OF ZONE BEATEN BY 75% OF SHOTS FIRED FROM A MAXIM GUN.

Dispersion of Cone.	500	1,000	1,500	2,000	Yards Range.
DEPTH	150	70	60	50	Yards.
WIDTH	4	8	13	19	Feet.

PROBABLE ERRORS IN RANGING TO BE ALLOWED FOR WHEN DIRECTING FIRE.

Method of Ranging.	P.C. of Error.	Extent of Ground to be searched to overcome probable errors in Ranging.				
		500	1,000	1,500	2,000	Range.
Judging Distance.	15	150	300	450	600	Yards
Judging Distance Combined with "Key Ranges."	10	100	200	300	400	,,
Range Finding Instruments	5	50	100	150	200	,,

[SEE OVER

LEFT A pre-First World War trajectory table for the M1893 Maxim showing the area covered by the cone of fire and the resultant beaten zone. Modified versions were issued as more streamlined ammunition was introduced. *(Author)*

ABOVE The Director No 4. Originally an artillery instrument, it was officially employed by MG units from 1917. It was used to provide directions to the gunner who could set the firing co-ordinates by using his bar foresight, direction dial and clinometer. *(Richard Fisher)*

LEFT A 'Dial Sight, MG Mk VIII', introduced in April 1939. This ingenious instrument rendered the clinometer, level bubble and deflection bar redundant, as it incorporated each of these elements. Pre-1941 patterns were later modified and the range drum recalibrated for Mk VIIIz ammunition. It remained in service until the Vickers gun was phased out. *(Ian Durrant)*

had changed quite dramatically since 1918, from direct to indirect fire, so rangefinding had become increasingly complex. During combat on the Somme in 1916, it became clear that the guns could be profitably employed in much the same manner as artillery, laying down long-range barrage fire at targets that were normally beyond observation. Maps had to be used, but as ordinary sighting methods were ineffectual the gunners still needed to hit these distant targets; therefore increasingly sophisticated methods were brought in to replace the simple range tables.

A deflection bar foresight had been introduced that allowed gunners to compensate for settlement or movement of the gun, permitting the sights to be regularly checked for direction. Slide rules that incorporated safety angles, horizontal and wind scales were introduced in 1924, but in 1933 the Fire Director, No 9, was at last officially issued in place of the old No 4 pattern which had been in use since 1917, issued to the Royal Artillery.

This comprised an angle-of-sight instrument with a lensatic compass (a magnetic compass with a magnifying lens on it) to provide greater accuracy. It was able to calculate angles of sight and plot directions based on the terrain which could be cross-referenced with the specially produced 'Plotter, No I Mk I', which had a degree plate to establish the line of fire from the gun to target. These readings were then transferred by the gunners to their clinometers.

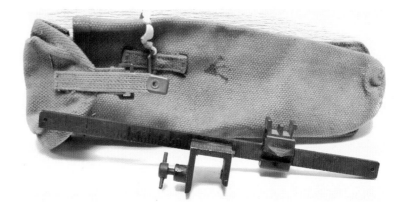

It was effective, but complicated to learn, and in 1939 an instrument often referred to as the 'crowning glory' of the Vickers gun was introduced. This was the 'Sight, Dial, MG Mk I'.

The dial sight was able to combine all of the former disparate elements and it was extremely versatile, with a lensatic sight containing a sighting triangle (not an optical aiming device as is often believed). It also incorporated a spirit level, range drum, angle-of-sight drum and deflection clicker ring, with ten minute-of-angle graduations for left and right deflection settings. Mounted on a sturdy bracket on the upper left of the receiver, it was easy for the gunner to see and it was initially graduated for calculating in conjunction with Mk VII ammunition. After 1941, a Mk II sight was introduced for the new Mk VIIIz ammunition, and conversion tables were supplied if only Mk VII ammunition was available. Both the plotter and dial sight stayed in service for the entire life of the Vickers gun.

ABOVE The 'Bar, Foresight, MG Mk I'. When clamped to the foresight of the gun, it allowed controlled changes of direction to be made when a fixed aiming mark was used, normally in conjunction with the clinometer. *(Richard Fisher)*

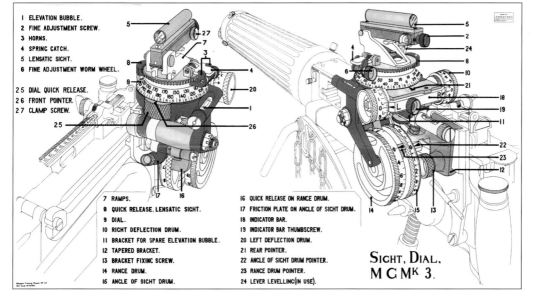

LEFT An instructional drawing showing the complex adjustment of the dial sight. *(Ian Durrant)*

Chapter Six

The Vickers in service

An overview of how the Vickers and Vickers-Maxim guns performed when used in the many roles for which it was adopted: with the infantry, on armoured fighting vehicles, at sea and in the air. In many instances the guns were successfully modified to perform tasks for which they had never been designed.

OPPOSITE A British Commando Vickers crew dig in and clean their gun. The gunner is holding the dismounted lock in his left hand. Photographed somewhere during the advance through Europe in 1945. *(Simon Dunstan)*

RIGHT Machine gun training was initially established for officers and NCOs only. This is the Hythe School of Small Arms in 1914. A group of officers are being inducted into the mysteries of the Vickers Mk I by a sergeant instructor. *(IWM Q53550)*

For land warfare between 1914 and 1918, the guns were employed by the infantry, cavalry and the new Tank Corps, as well as armoured car units and motorcycle units. Each branch possessed slightly different forms of organisation but training for gunners remained the same throughout the war. Initially, machine gunners were an integral part of their regiments working at battalion levels, so for example in 1914 the 1st Battalion Royal Fusiliers would have had one Vickers-Maxim section comprising of a mere two guns, with sixteen men, two NCOs and an officer commanding, and within each brigade there would be four machine gun sections.

Training on the Vickers was undertaken at the Machine Gun School at Hythe in Kent, but only for officers and NCOs, who were thought capable of training the modest numbers of machine gunners required. By mid-1915 it had become very apparent that both the number of guns in service and the training were quite inadequate. The huge increase in production requested by the Army meant that there had to be a corresponding increase in men being trained. Fortunately, there were some senior officers who believed passionately in the future of the machine gun, the most significant being the delightfully named Brigadier General Christopher D'Arcy Bloomfield Saltern Baker-Carr, who had also been a musketry instructor at Hythe. Due mostly to his persistent lobbying, the Machine Gun Corps was formed by Army Order on 22 October 1915. From its modest

RIGHT An early war photo of a section of the 2nd Coldstream Guards Machine Gun section with two M1893 Maxims and an early rangefinder. The two gun-carrying limbers can be seen behind them. *(Courtesy Pete Smith)*

beginnings it would soon consist of Infantry Machine Gun Companies, Cavalry Machine Gun Squadrons and Motor Machine Gun Batteries, and by 1918 it had a complement of 170,000 officers and men, suffering 62,000 casualties during the war, earning it the well-deserved soubriquet 'The Suicide Club'.

A machine gun training school had been started at Wisques in France in 1914, but was soon inundated by the numbers of men required, as training was no longer the preserve of officers and NCOs, but the neophyte gunners as well. An Overseas Base Depot was established at Camiers in the Pas-de-Calais and men who already served in Battalion MG sections were automatically transferred to the MGC, while volunteers were called for to help fill the ranks. There was no shortage. Perhaps in part this may have been because when in the front lines, MGC crews manned their guns 24 hours a day, 7 days a week, on a rota system. They did not have to undertake the more mundane work expected of infantrymen, such as fatigues, patrols, sentry duties and so on. Each MG section worked as a close-knit team, forming strong bonds.

According to Army regulations a machine gun crew consisted of seven men. No 1 was the gunner, who was responsible for the cleaning and proper maintenance of the gun. He carried the tripod into action, often wearing the unique padded waistcoat issued to crews to help spread the weight of the tripod legs over the shoulders. He would then assist his No 2 in setting up the gun. He was responsible for carrying out fire orders, observing the effects of his fire and making any required corrections. His No 2 carried the gun and mounts and ensured a regular supply of ammunition for the weapon. He also watched for signals from the MG officer, and relayed them to the gunner. Nos 3 and 4 were ammunition carriers, who moved the loaded boxes from the dump to the gun position. In theory No 3 was also responsible for the water supply, but either man could take responsibility for this as circumstances dictated. No 5 was the scout, who would liaise between the Section Officer and gun team, the officer possibly being in a location out of sight. No 6 was the range-taker, who checked range cards and took readings. He would also hold any maps to ensure correct lines of

ABOVE Men of the 3/5th and 3/6th Battalions, Royal Norfolk Regiment, sit behind their .303 Maxim. An early No 3 Mk I ammunition box appears to be in use. By 1915 these men would have been transferred to the newly formed Machine Gun Corps. *(Courtesy Pete Smith)*

LEFT An unusual photo showing a Vickers with anti-aircraft sights apparently engaging a low-flying aircraft. However, the plane appears to be a British RE8, which was doubtless flying for the cameraman, or perhaps acting as a target-towing machine. *(Ian Durrant)*

RIGHT MG waistcoat, gloves and facemask.
(Richard Fisher)

fire if the target was not under observation. No 7 was the 'spare bod', often employed to reload belts if ammunition was running short. It should be stressed that every man in the crew was a trained gunner, who could perform all of the functions of the others. In the light of their combat casualties, a gun crew could often be reduced to two or three men. The Section Officer was responsible for siting the guns, issuing fire orders and ensuring any map references were correctly adhered to, as well as passing on orders received from the Brigade Machine Gun Commander.

Of all the MGC units raised in the First World War, by far the largest was the MGC Infantry, who were allocated into four 4-gun sections, each of which made up a 16-gun company.

They were then placed into Brigade Machine Gun Companies, initially three per division, but demand caused this to be raised to four in 1917. Thus, each division had 48 guns for its 12 line battalions. To simplify command in March 1918, the four companies of each division were formed into individual Machine Gun Battalions, which were given a battalion number. Of the infantry MGC units, only the Guards Division held the unique honour of having its own Machine Gun Companies, formed into the Guards Machine Gun Battalion, in 1918 becoming the Guards Machine Gun Regiment who wore a separate cap badge. The Cavalry Branch consisted of Machine Gun Squadrons, one per cavalry brigade and there was also the Motor Service. Initially this was formed from the Royal Field Artillery and became the Motor Machine Gun Service (MMGS), mounted on Clyno motorcycles. When the MGC was formed, the MMGS became the Motor Branch, which diversified into several different units: motorcycle batteries, light armoured motor batteries and light car patrols. As well as motorcycles, other vehicles included Rolls-Royce and Ford Model T cars. The Heavy Section, MGC, was formed in March 1916, becoming the Heavy Branch in November, and these men were to form the crews who would man the first tanks ever to go into action at Flers on the Somme on 16 September 1916. In July 1917 the Heavy

RIGHT Cheerful men of the MMGS on the Somme, late 1916. Their Clyno motorcycles have sidecars with Mk I Vickers guns and many of the crew wear captured German headdress. Some 478 Clynos served on the Western Front.
(IWM Q6231)

LEFT Warmly clad recruits at the MGC training centre, Belton Park, watch as a Vickers is fired. *(Author)*

Branch, by then much expanded, separated from the MGC to become the Tank Corps, later called the Royal Tank Regiment.

By 1915, the requirement for an almost endless supply of Vickers guns and men led to the formation of a large training establishment in England. In spring of that year Lord Brownlow offered Belton Park in Grantham, Lincolnshire, as the new MGC training centre. Organising an entirely new regiment sourced from regulars, territorials and volunteers was a Herculean task, as Lt Col Seton-Hutchinson wrote:

Almost at once thousands of men began to pour into the wooden huts which rapidly spread themselves over Lord Brownlow's parklands at Belton Park. Thousands of horses, mules, and vehicles appeared; and, with two weeks of wintry rain, the park was submerged beneath a sea of mud. The task of sorting and re-equipping all conditions of men, in every kind of uniform, some holding the rank of sergeant and corporal, from the various New (Kitchener) Army battalions from which they had been drafted, other regulars and special reserve soldiers from regimental depots with much machine gun experience, would have tried the patience of a Job.

Here a thorough ten-week course covering every possible eventuality was held for both officers and men, although the syllabus differed between them. For officers it concentrated more on fire control, training, tactics, organisation and command. For NCOs and men, it was more 'hands-on', with emphasis on care and cleaning, gun stoppages and immediate action, belt-filling, target indication and recognition, understanding range cards, sighting devices, ground cover and camouflage. All were thoroughly covered by experienced NCO instructors in a course that ran eight hours per day, six days a week. A final examination had to be passed for a man to be given the title 'machine gunner', or 'Emma Gee' in Army phraseology, and wear the coveted crossed Vickers gun cap badge.

At the outbreak of the war, the initial employment of machine guns had been for their firepower to help repel attacks, the guns being set up in front of, or within, the trench lines. Except in dire emergency they were seldom used to provide point-blank frontal fire,

BELOW A rare front-line image of an early five-arch Vickers and gun team purportedly under fire in the Ypres sector in early 1915. The man nearest the camera has an ammunition box at the ready. *(Author)*

ABOVE The MG section of the 1st Rifle Brigade, 1914, with Cpl J.W. Brooks at the front, far left. He was to win both the MM and DCM as a machine gunner. It is interesting to note that both guns have been camouflaged. *(Eugenie Brooks)*

BELOW A Mk IV Tripod Direction Dial. The degrees are marked from 0 to 360 and enabled the gunner to precisely lay the line of fire according to the co-ordinates given to him by the commander. *(Ian Durrant)*

but were set up at an angle to the front line, with interlocking fields of fire through which any advancing enemy would have to pass. This was referred to as enfilade fire. To do this the range had to be correctly estimated, using a Barr and Stroud rangefinder; the direction was then taken by compass, the precise angle of which was given to the gunner by the fire commander. This was in degrees and corresponded to the circular brass direction dial set on top of the tripod socket; the elevation of the barrel was measured in degrees and minutes, the trajectory of the type of ammunition also having to be calculated from the range table. The unit's clinometer was then used to set the precise barrel angle, normally by the officer commanding the section. The gunner ensured that the tripod legs were securely fastened using the three locking levers, which had to be periodically checked to ensure they did not work loose and alter the direction of fire. Wherever possible, the legs would be weighted down with sandbags, which helped to steady the gun, ironed out some of the vibration of firing and reduced the likelihood of the weapon shifting.

Direct or enfilade fire was the keystone to early machine gun tactics and when properly applied was devastating – the only defence was to try to knock out the guns with artillery. Traversing a gun was done by 'tapping' with the heel of the palm of the hand. Experienced gunners could move the arc of fire 1° per tap, which would either bring it back on to target or alter the dispersion of bullets within the cone of fire if being fired at long range. The ideal type of fire to use against a visible attacking force was to aim 'grazing fire' at roughly knee height, so that men were felled, dropping into the oncoming bullets, invariably being hit for a second or third time. Even if the range judgement was over or under, such fire would result in the feet being struck (equally effective) or in upper body hits.

There were some problems when employing close-range tactics, though, for the Vickers was hard to move once dug in, vulnerable to enemy artillery fire and easily disabled by bullets, shrapnel or shell splinters. Ideally, guns needed to be sited further back, out of direct sight of the enemy, from where they could still pour heavy fire down on to the lines of communication, assembly areas and roads. Consequently, the numbers of Vickers guns kept in the front lines was reduced and as time progressed, they became more commonly found in reserve trenches, or situated on any nearby high ground to the rear.

To engage an enemy with indirect fire effectively required considerable planning as the target was not in sight and could only be plotted with the aid of maps. In fact, getting any Vickers guns into the line was always a major consideration, as road transport could seldom take them closer than perhaps a mile, from where everything had to be hand carried by the crews. An efficient gun-carrying harness for pack animals had been devised prior to the war but the problem with trench warfare was that, like vehicles, horses or mules were limited as to how close to the front they could be brought, although the pack equipment was much employed in the Middle East, Salonika and Mesopotamia. But on the Western Front, certainly through the stagnant years of 1915–17, most of the equipment required had to be carried by the gunners themselves. This was not too onerous for young fit men, but if the guns were to be used behind the lines for barrage firing, then vast quantities of ammunition were needed, and in this instance wagons would be brought as close as possible, and grumbling infantry carrying fatigue parties would be ordered to help move the boxes. Generally each gun was allocated 3,500 rounds, although this was often exceeded by several thousand rounds depending on the nature of the barrage. Indeed, one of the major logistical challenges that emerged from the war was the sheer physical effort involved in getting the guns into the firing lines and keeping their

LEFT Some concept of the quantity of ammunition required by a gun section can be gleaned from this photo of a New Zealand MG unit conferring by a stockpile of about 100,000 rounds of ammunition. This would enable four guns to fire an overnight barrage. *(Wellington War History photo, Turnbull Library G-133211/2)*

ABOVE Two views of the 'Pack, Horse, Vickers .303 inch Machine Gun'. The first image shows the gun strapped in place, with the Sangster tripod mounted upside-down to enable it to fit properly. On the right side of the horse are six ammunition boxes. *(IWM Q035658)*

ABOVE RIGHT The second horse carries the tripod, spares boxes as well as ammunition. Gunners took great pride in their horses and disliked their inevitable replacement by vehicles. *(IWM Q035659)*

voracious appetites for ammunition, water and barrels fed.

The first time properly co-ordinated indirect fire was employed was during the Battle of Loos, on 9 September 1915, when the guns of the two London Divisions opened a continuous overhead barrage on German assembly areas, preventing reinforcements from being brought up to the lines. It was the first recorded instance where machine guns mimicked the tactics of the artillery, although sadly it made little difference to the eventual outcome of the battle.

As the war progressed it became increasingly commonplace to engage targets that, as far as the gunners were concerned, existed only on maps. Setting up a gun was the same routine, except that firing over one's own troops was fraught with danger, for if a gun worked loose on its mount or the tripod legs dropped slightly then a barrage of .303 bullets would drop into the backs of unsuspecting soldiers. Very careful placement and weighting of the tripods was done, and extremely precise measurements made of the distances and angles of fire to ensure that the guns were perfectly laid on the target. It was not high-precision shooting, but it didn't need to be, trenches were static and if the target was on a road or in a wood then it was unlikely to be moving at greater than a walking pace.

As they descended, the bullets created what was known as a beaten zone. This is the zone into which the falling bullets will descend, creating a cone of fire. It is caused by the positive and negative jump of the barrel as it fires, each shot leaving the gun alternately high and low as the recoil fractionally moves the muzzle. It is a relatively small movement, but over 500yd it will create a beaten zone over 150yd long and 4ft wide. The greater the distance, the shorter and broader this becomes, so that at 2,000yd it will be 55yd long, but 19ft wide. At Cambrai on 21 November 1917, the 3rd Army had no fewer than 100 guns in place to give supporting fire for the attack on Bullecourt. All the Vickers guns opened up, each firing at 500r/pm and the subsequent barrage of 800 bullets per second utterly laid waste to areas occupied by German reserves, and was substantially responsible for the early success of the attack.

As the fighting began to move from static to a war of movement after 1917, machine gun tactics had to be adapted. Defence in depth became the watchword for the retreating Germans with their machine gun units sited in small pockets along the lines, providing defensive and harassing fire, materially holding up the advancing Allies. In order to combat this, MGC scouts observed enemy positions, and reported back, enabling accurate covering fire to be given to the advancing troops.

ABOVE LEFT A gunner of the 2nd Argyll and Sutherland Highlanders at Bois Grenier has set up his Vickers for anti-aircraft use by the simple expedient of raising the tripod legs on sandbags, early summer 1915. *(IWM Q48967)*

ABOVE A M1893 Maxim with brass handles and rolled feedblock is used to instruct members of the machine gun section of the 1st Battalion, Cameronians (Scottish Rifles) at Grande Flamengrie, Bois Grenier, February 1915. *(IWM Q51584)*

At Amiens on 8 August 1918, a huge, co-ordinated Allied attack was launched against the Hindenburg Line, the beginning of the Hundred Days Offensive. Working in close co-operation with the Royal Artillery, machine gun units across the front laid down an advancing barrage *ahead* of the artillery's creeping barrage to prevent German troops from evacuating their trenches, or reserves from coming up. The tactic worked brilliantly, effectively catching the enemy between a rock and a hard place and enabling the line to be captured.

Although not appreciated at the time, this was to prove the end of the line for the MGC, for it had become clear that the issue of lighter, more portable machine guns such as the German MG08/15 and the Allies' Lewis gun was the way forward. Faster to manufacture, easier to carry into battle, requiring fewer crew and training, they were also less vulnerable and clearly made more tactical and economic sense. But the British Government preferred to believe that no future war on such a scale was ever likely to occur again. As a result of a series of drastic cuts the Army was reduced in size, the Machine Gun Corps disbanded in 1922 and with them all the skills and experience so hard learned were lost. The Vickers guns once more returned to the care of infantry battalions and cavalry regiments.

If there was one area where the Vickers proved to be quite unexpectedly successful, it was in an arena that would have been impossible to imagine in 1914: the air. It is a surprising fact that, of all the Vickers guns manufactured during the First World War, a third

RIGHT Motor machine gunners engage a German aircraft near Ypres, March 1915. The spotter at the right is following the target with the 20× power field service telescope. Their 552cc Scott motorcycle was underpowered and struggled with the load it was required to carry. *(IWM Q61581)*

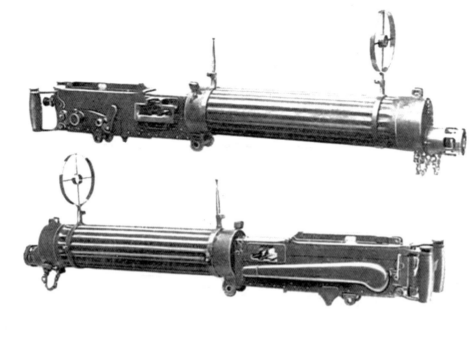

Mk I Gun seen from the right

Mk I Gun seen from the left

Mk I Gun seen from above

Mk II Gun Arial seen from the right

Mk II Gun Arial seen from the left

Mk II Gun Arial seen from above

OPPOSITE A rare page from a 1918 report on air service machine guns. The top gun is no more than a ground Mk I Vickers with a ring and bead sight fitted and no modifications to the mechanism or cooling system. The lower three images show the Mk II, with cooling louvres, open fusee spring and modified firing mechanism to permit synchronisation. *(Author)*

ABOVE An RAF air gunnery range somewhere in France. The pilot sits in a mock-up cockpit. The target was usually railtrack-mounted to provide practice in deflection shooting. The officers on the right have been tasked with filling ammunition belts. *(IWM Q010369)*

LEFT Mk I Vickers converted for air training use showing the Hyland extended loading lever, and early Mk I steel ammunition links. The base of the clamped-on bead sight is just visible on the top of the water jacket. The cocking handle position indicates a No 3 stoppage. *(Author)*

were produced solely for air service. Like tanks, aerial warfare was a phenomenon of the First World War that arose from nowhere.

The initial employment of aircraft was simply for observation, the pilots of the opposing sides taking pot-shots at each other with pistols or rifles. If they carried an observer then it was simply a matter of time before he became armed as a defensive measure. But the design of early aircraft meant that there was no method of attaching a forward-firing machine gun without shooting off the propeller. A machine gun could only be carried if the engine was of the pusher type, mounted behind the pilot. Besides, it was hardly an ideal environment for a belt-fed weapon such as the Vickers, for while the slipstream kept the water in the jacket cool, the ammunition box had to be mounted somewhere, and aerial manoeuvring could simply cast the belt adrift, where it flapped dangerously in the slipstream, shedding cartridges. Another unforeseen problem was that the Vickers ejected its empty cases downward, filling the cockpit floor with loose brass which was potentially hazardous and capable of jamming controls. The more widespread introduction of the Lewis by the beginning of 1915 proved fortuitous, and it became widely adopted as Britain's secondary aircraft machine gun, predominantly for observers or mounted on the upper wing of an aircraft, above the propeller.

However, the new breed of fighter aircraft emerging in 1915 required a primary weapon, and the need became more pressing in the wake of the successful introduction of the Fokker-Lubbe system, which permitted a Maxim gun to fire through the arc of the propeller in synchronisation with its rotation. There was a downside associated with this, for the gun could fire only in staccato bursts, at an impractically slow rate of fire of around 350r/pm, in order that the bullets missed the blades. This was simply not fast enough to guarantee hits on the fleeting targets that were presented in air combat. Nevertheless, by the early summer of 1915 the RFC were desperately searching for a solution.

Guns such as the Lewis and Hotchkiss were unsuitable in a forward-firing role, as they worked on an open-bolt system and could not be timed to fire with any form of interrupter gear. The closed-bolt Vickers was the clear candidate, but how could it be adapted? As usual the ingenuity of several British engineering companies came up with several devices; Airco, Armstrong Whitworth, Martinsyde, Sopwith-Kauper and Vickers-Challenger all produced variations on a theme, utilising cams attached to a suitable engine component, such as the oil pump, camshaft or special cams fitted to

BELOW Some of the early types of ammunition belts tried in aircraft Vickers guns. The webbing belt was never satisfactory, but finding a steel-linked pattern that functioned reliably was a difficult process. The Mk II was a copy of a German type, but was abandoned in favour of the Prideaux pattern. The Mk III became the standard for British aircraft. *(Author)*

the rear of the rotary engines. These were connected to the Vickers by means of shafts or cables, initiating fire when the propeller was not in line with the muzzle, but high rates of fire still could not be achieved.

The Vickers-Challenger had proved relatively effective, but in late 1916 Major L.V. Blacker, the Vickers engineer who had worked on it, heard of a new hydraulic synchronisation system that had been designed by an ex-patriate Romanian engineer called George Constantinescu. Vickers speedily arranged for the drawings and an early prototype to be delivered to Blacker.

Worked by means of gears attached to the main propeller shaft, the revolutions created impulses via a hydraulic generator pump that was linked directly by a hydraulic line to the machine gun's trigger. The pilot had full control of the mechanism and only needed to prime the system with a small hand pump, which he could disable if necessary, simply by moving a lever from 'Fire' to 'Safe'. Because it was not mechanical it enabled a Vickers gun to be fired at far higher speeds than any existing system. In fact, the hydraulic pressure created was actually capable of firing the gun at a far higher rate than was safe for it, and this raised a number of secondary problems that required solving.

The rate of fire needed to be kept down to 800–900r/pm, and the solution to controlling this was found in the relatively simple invention of a naval engineer, Lt Cdr George Hazleton. The Hazleton device comprised an extremely strong buffer spring and a vented muzzle booster that replaced the standard unit, and it harnessed a greater proportion of the combustive gas, raising the rate of fire up to 1,000r/pm. The problem was that any sustained fire over the 500r/pm the gun was designed for merely placed undue stress on the mechanism. Although strong, the internal components of the Vickers were not manufactured to cope long term with such high rates of fire, and some parts would clearly require reworking. Then there were the questions of cooling, the amount of ammunition to be carried, the method of cocking the gun and dealing with the inevitable stoppages.

The problems were gradually resolved; the Hazleton was modified with larger vents to permit more gas to escape, thus slowing down the cyclic rate to an ideal 850–900r/pm, and a shorter, lighter firing pin was produced with a stronger firing pin spring. In late 1917 a fusee spring with a different type of tempering was fitted along with an external linkage that enabled the pilot to adjust it from the cockpit. Vickers introduced their own modifications as well; these included a deeper, larger muzzle cup, a buffer spring behind the crankshaft to cushion the recoil action and a stronger and more efficient linkage for the fusee spring. Fortunately, the cooling question was taken care of by the fact that the guns were carried in the 100mph slipstream of the aircraft. Initially letting this simply cool the water in the jacket progressed to dispensing with the water entirely and opening up the jacket by means of drilling large entry holes in the front endcap to direct the air

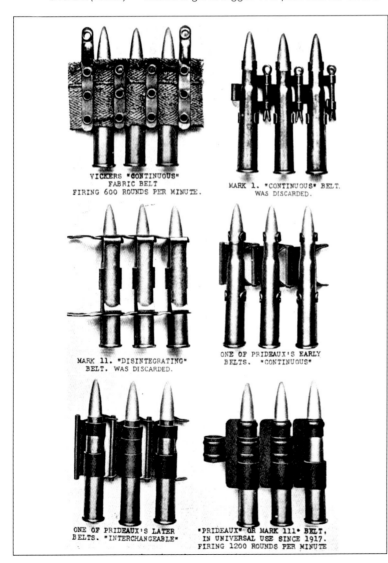

through, as well as cutting side louvres or gills in the rear of the jacket.

Cocking the gun had originally been done by a number of types of enlarged cocking handles designed to be grasped by a heavily gloved hand but, should a stoppage occur, simply working the cocking lever would not always clear it. The normal remedy was to open the top cover of the gun and manually clear the jam. Of course, this was a nearly impossible task in air combat and the reason so many dogfights were broken off; a jammed gun resulted in a defenceless pilot streaking for home. A solution appeared in the spring of 1917 in the shape of the Hyland Cocking System. This had the usual extended wooden-handled cocking lever, but it served a dual function; pulling back on it did not merely cock the gun, but in the case of a misfire a second pull drew back the barrel, lock and recoil plates, automatically pulling the belt through the feedblock, ejecting the cartridge and clearing the jam.

There then remained one crucial issue to be dealt with, and that was the ammunition supply to the gun(s). As previously mentioned, cloth belts were a disaster in the air; indeed, several pilots had been injured by empty belts whipping into their faces. In the wet, the belts often failed to feed properly, and the answer seemed to lie in finding some form of expendable metal link that was simply ejected along with the empty cases. Fortuitously, the answer came from the Germans when in late spring 1915 an Albatros D1 scout crash-landed intact behind British lines and was taken for evaluation by the RFC. Its forward-firing Maxim contained a type of disintegrating steel link, and Britain immediately copied it, though in service it proved rather problematical.

The cartridge was inserted into a long metal tube, with two thin sideplates that then located over the next cartridge that had to be

ABOVE The Bristol Fighter was one of the best two-seat all-rounders produced during the First World War. The pilot's single Vickers with the Hyland cocking lever is seen just above the engine nacelle. *(Author)*

inserted. If the links and holder were not made to very exact tolerances, or if they became damaged, the cartridges jammed in them and did not extract; neither did the links always pass smoothly through the feed pawls of the Vickers gun. RFC armourers spent an inordinate amount of time loading belts and testing each link to ensure the cartridge fitted perfectly as the life of their pilot depended on it.

In late December a solution emerged from the unlikely workshop of a Surrey dentist named William de Courcy Prideaux. He had long been interested in loading devices and had perfected a speed-loader for the .455 Webley revolver. In late 1916 he came up with a stamped rolled link into which a cartridge was inserted and held firmly by a central clip

BELOW An early British conversion of a Mk I ground Vickers for air service. The rear sight has been removed, a strengthened base mount has been added and cooling louvres cut in the jacket. It retains its firing grips, trigger and crank handle. *(IWM FIR 9193)*

ABOVE American fighter ace Eddy Rickenbacker in the cockpit of his Nieuport 28 fighter, 1918. He has a pair of Colt-Vickers guns, one on the nacelle, the other fuselage mounted. These were each 8lb lighter than the Vickers ground guns. *(US Signal Corps)*

ABOVE This image of an RE8 shows a Vickers operated by the hydraulic Constantinescu system. One problem with this arrangement was the gun's tendency to freeze in cold weather. Sighting was by means of an Aldis optical sight visible in front of the pilot. Note the aircraft pattern Lewis gun on the observer's Scarff ring. *(Author)*

RIGHT The cockpit of a Sopwith Dolphin, with its twin Vickers guns. At centre the left and right feedblocks can be seen, and the aluminium belt guides sit in between them. The Aldis sight is mounted over the airspeed indicator. Both guns have cheek-pads to protect the pilot in case of a crash landing or violent manoeuvring. *(IWM Q68704)*

RIGHT An interesting sectional image of the front of a Sopwith Snipe, introduced in late 1917, with its twin nacelle-mounted Mk II Vickers. One ammunition container and its hinged lid can be seen in the centre of the photo. Both guns sit on top of the fuel tank! *(IWM Q69712)*

that had two internal ribs, with two other clips, upper and lower on the next link providing free movement to the belt. They were also precisely manufactured to fit the feedblock of a Vickers gun. The Prideaux belt was immediately adopted for service and produced in vast quantities. Mitchell & Sons of Birmingham, one of the main contractors, manufactured almost 72,500,000. Unlike ground service ammunition belts, they could be joined to form lengths greater than 250 (often up to 400 or 500 rounds) but they couldn't be reused and were thus considered disposable. Sadly, Prideaux was never to receive a penny in royalties for his invention from the British Government, and he died in 1922. All of these improvements contributed to a gun that, by the beginning of 1918, was not only firing reliably but also at nearly twice the rate of the mechanically synchronised German Maxims, providing Allied pilots with crucial firepower superiority.

It was not until the open warfare began on the Western Front in 1918 that the much underemployed motorcycle sidecar outfits and light trucks of the MMGS eventually came into their own, having been brought back to front-line service once the stalemate of trenches had been overcome. They soon began to prove their worth by being able to move quickly into enemy territory and wreak havoc, as well as providing vital mobile support for attacking infantry units.

While the motorcycle units did not always fare so well in the desert conditions of Palestine and Egypt due to heat and sand, the armoured cars did better, often covering huge distances

LEFT A Model 1914 Rolls-Royce armoured car, with the turret-mounted Vickers gun under its armoured cover. Although impractical on the Western Front, they saw considerable service in Africa, the Middle East and Russia. It weighed 4.7 tons and had a top speed of 45mph. *(Author)*

ABOVE A Mk I female tank of the Heavy Branch, MMGS, photographed in September 1916. The four Vickers guns were sponson-mounted in pairs and needed armoured protection for their vulnerable jackets. This proved impractical and they were soon replaced by the Lewis gun.
(Tank Museum)

to launch attacks on Turkish positions when they were least expected. The Vickers was the most widely used vehicle weapon, although its water jacket inevitably made it vulnerable to small-arms fire. These were to be the forerunners of later units such as the Long Range Desert Group, where surprise, speed and firepower were of the essence.

The Vickers was also initially supplied to the neophyte Heavy Branch, Machine Gun Corps, whose lumbering tanks went into action on the Somme in September 1916. The female tanks were equipped with four Vickers guns covered with armoured jackets to offer protection against enemy fire, but they proved unsuitable from the outset. Tank crews begged to be able to have the Lewis or French Hotchkiss guns, but they were overruled, so the first tank attack in history was accompanied by the distinctive rattle of Vickers guns. Naturally, once the water was lost, the gun overheated and was soon rendered unserviceable as it was impossible for a gunner to leave the protection of the vehicle to replace the barrel or effect repairs. This was to become a recurring theme with machine guns for armoured fighting vehicles (AFVs).

Supplying ammunition was difficult too, for a metal frame had to be hooked to the gun for the ammunition box to sit in, which hindered its movement within the sponson's ball

RIGHT A Royal Naval Air Service armoured Seabrook lorry, mounting a M1893 Vickers-Maxim and Hotchkiss 6-pounder QF gun. The machine gunner's close proximity to the Hotchkiss would have been deeply unpleasant when it was fired.
(Tank Museum)

mounting. In addition, huge quantities of belted ammunition (25,000 rounds per tank) had to be squeezed into an already overcrowded space. After the Somme, most tanks were equipped with an altered mount into which the ubiquitous Lewis gun fitted, which proved a far more acceptable weapon. Even when its jacket was holed, the Lewis gun would still function quite effectively and its drum could easily be reloaded from boxes of loose ammunition. By the end of the war, the tank had overcome initial military scepticism and become an integral part of the British Army. Unlike the disbanded MGC, which had unreasonably been seen as simply a wartime expedient regiment, the same fate did not affect the Tank Corps, which in October 1923 officially became the Royal Tank Corps. Trying to harness the firepower of the Vickers was to become an ongoing problem for them in the future.

With the end of hostilities in 1918, and the subsequent disbanding of the MGC, initially there seemed little future for the Vickers gun. However, as another war became likely, Britain began to quietly reorganise, equip and train their armed forces, determined not be caught unprepared for a second time. Initially each infantry battalion or cavalry regiment had its own MG platoon, later a whole company; however, as mechanisation developed it was decided that five Regular Army battalions would be designated MG units, these being the Middlesex Regiment (six battalions), the Manchester Regiment (five battalions), the Royal Northumberland Fusiliers (five battalions), the Cheshire Regiment (five battalions) and the Kensington Regiment (Princess Louise's – two battalions). One battalion each of the Argyll and Sutherland Highlanders and the Gordon Highlanders were designated to support the Scottish divisions but these were disbanded in 1940, as was an MG battalion of the Devonshire Regiment. MG units were smaller than in

1914–18, each company having three, not four, platoons, so a battalion complement was 48 Vickers.

When war began in 1939, regimental depots were once again unable to cope with the expansion of machine-gunner training so two centres, Saighton and Blacon, were established in Cheshire. Many ex-machine gunners from the First World War were made NCO instructors, several of whom were sent to train at MG school at Netheravon in Wiltshire. It says something about the pride that these men had in being Vickers gunners that at no time during the war did the training camps suffer from a shortage of instructors.

ABOVE The reason why the water-cooled machine gun was not a practical proposition in an armoured vehicle is evident here. The Maxim mounted on this Lanchester armoured car has been hit by shellfire. *(IWM Q72874)*

RIGHT Although much reduced, the training of Vickers gunners continued through the 1920s. These immaculately turned out soldiers of the Middlesex Regiment are practising their section fire drill on a playing field. The gun is a post-1918 variant with plain steel jacket and armoured nose cone. *(Author)*

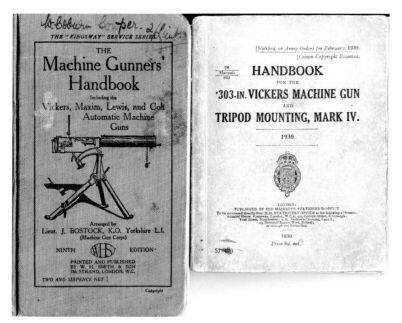

LEFT Two training manuals, 15 years apart. The left handbook was commercially produced in 1915, as no official Vickers manual was available. It was owned by 2/Lt W. Cowper, 1st Surrey Rifles and MGC. The other handbook is the first official one printed by His Majesty's Stationery Office in 1930 and was issued to the Canadian Saskatoon Light Infantry MG section. *(Author)*

Training was shorter and slightly different, for the war was going to be far more fluid, and post-1940 it became clear that machine guns would rarely be sited as front-line defence weapons, because they were now simply too vulnerable, not only from artillery and tanks, but from the increasing use of mortars. The most widely found German mortar, the 81mm Grenatenwerfer 34, introduced in 1934, had a maximum range of 1½ miles and could be fired from deep cover invisible to observation, dropping its bombs vertically into concealed MG positions. So the Vickers became predominantly a weapon to provide indirect and overhead support fire and gunners were given detailed instruction in these fire-control methods.

In terms of time and training costs, it was believed that training gunners to master the complexities of rangefinding apparatus was largely wasted as, once a gun was set up, it rarely required much alteration. Besides, the now commonplace issue of platoon radios meant that fire-control orders or amendments could be made far more easily. So, from September 1944, MG range-takers were still employed but guns were being used as pairs far more frequently, so they were held at a two-gun section level under control of the Section Commander rather than with individual gun teams. The range-takers were trained as a separate unit from the MG companies and were allocated where needed to provide precise data to MG units as they set up, being then rapidly transferred to other units as and when required.

After the war ended in September 1945, there was yet again a great cull of the machine gun units, with the machine gun battalions being disbanded and MG companies reduced to six gun platoons, one per battalion. The last newly manufactured Vickers gun was supplied to the British Army on 1 June 1945. In 1947 the Vickers was withdrawn from Regular Army service, being retained only by Territorial units. This may well have signified the end of its service life, but world events were to prove otherwise.

The Cold War and the looming threat of Communism, as well as expanding Chinese influence, began to worry the governments of the US and Europe, and in 1950 the Support Weapons Wing, formerly at Netheravon, was

BELOW Many ex-MGC men volunteered for Home Guard service during the Second World War. Here two mature gunners pose on the village green with an American-supplied M1915 Colt Vickers. Most of these were chambered for .30-06in ammunition and had a red stripe painted around the jacket. *(Author)*

re-established – just in time as it turned out – for war broke out across the disputed North/South Korean border on 25 June 1950. In many ways, Korea's mountainous terrain and vast expanses of open fields were the perfect setting for the Vickers, which excelled in its role in providing long-range supporting fire. In particular guns manned by the Australian Army were widely employed to lay down murderously effective fire during mass attacks by Chinese and Korean troops. Earlier problems caused by winter weather were no longer as serious as they had been, for by now anti-freeze was widely available. Added to the coolant in the jacket it prevented the guns having to be drained or regularly fired, although most crews preferred to keep the guns functioning in very cold weather by shooting off short bursts as their predecessors had done.

By the cessation of the hostilities in 1953, it was obvious to everyone that the new breeds of mortars and the increasing availability of the heavy machine gun, predominantly the .50-calibre M2 Browning, were relegating the water-cooled medium machine gun to history. In 1957, after exhaustive testing, the British Army had adopted the FN MAG in the guise of the L7 General Purpose Machine Gun, a gas-operated, open-bolt light machine gun based on the German MG42, and it was chambered for the newly adopted 7.62 × 51mm NATO cartridge.

But it was not quite the end of the road for the Vickers gun, for it was still to see service in peripheral conflicts around the world as Britain fought small wars and undertook policing operations in places such as Malaya, Oman, Cyprus and Aden. Indeed, it was in Aden that the Vickers made its final appearance, being used in support of operations at Radfan, about 30 miles north of Aden, in May 1964. On 30 March 1968 it was declared obsolete and was officially withdrawn from service.

The question for the Ministry of Defence was what to do with the thousands of stored Vickers guns at Army depots such as Weedon. It was certainly possible to convert them to the new NATO calibre, but this meant manufacturing replacement barrels, as well as some internal modifications and a new pattern of ammunition belt, the costs of which would be prohibitive. Exactly how many were stored is a moot point, but the number was probably in the region of 45,000 – and no one wanted them. In fact, a mystery still surrounds the fate of the surviving guns, which were not passed into the commercial collectors' market or apparently sold off to other countries. One ex-armourer who spoke to the author some years ago recalled in the late 1960s packing dozens of guns into metal crates which were then dumped in deep water off the English coast, so this may explain the rarity today of any British-made Vickers guns.

ABOVE A Barr and Stroud rangefinder being used with a foxhole-emplaced Vickers. The shallow angle of the gun and the fact the gunner is employing the rearsight would indicate the target is not far off. *(Author)*

BELOW A very youthful pair of British soldiers fire their Vickers in support during the Kenyan uprising of 1954. The picture illustrates how ejected cases from the gun always drop vertically from it. *(Simon Dunstan)*

Chapter Seven

Vickers variants

Thirty countries, excluding the British Dominions, adopted the Vickers gun for military service and this chapter takes a look at some of the variants employed, as well as officially sanctioned copies, such as the American Vickers Model 1915 and the unofficial copies made in huge numbers by Russia as their Model 1910 guns.

OPPOSITE A rare Argentinian export model World Standard Maxim, one of a batch manufactured to chamber the 7.65mm cartridge. Note the elegant but complex elevation and traverse mechanisms and the bicycle-type seat. *(Rock Island Arsenal)*

ABOVE A Mk. IVB Vickers .303in tank machine gun, the first pattern to be adopted for AFV use. *(H. Woodend/NFC)*

BELOW The first .5in AFV Vickers gun. In most respects, it was very similar to the .303 guns, though it did not require a muzzle booster. *(H. Woodend/NFC)*

The largest numbers of variant Vickers guns were those manufactured for AFVs, totalling eight different models chambered for .303in ammunition, all of which were specifically designed for vehicles. Indeed, they were virtually unusable as infantry weapons, having no provision for tripod mounting and weighing between 32 and 49lb each. As with the air service guns, initial ideas for simply converting surplus First World War-era guns proved totally impractical, for many reasons. Space was at a premium in a tank, ammunition feed was difficult, they could not be fitted securely into turret ball mountings (as bullets could penetrate the jacket and interior of the vehicle), the barrels and water jackets were too long and spade-grips were impractical. Then there was the problem of water-cooling. . . .

The guns that were produced ran from the Mk IVA of early 1930, simply a modified Mk I gun with the addition of a pistol grip, through to the final pattern, the more sophisticated Mk VII of 1936. This development of these guns over this six-year period was remarkably rapid, but most of the mechanical problems associated with the Vickers guns had been ironed out during the First World War, so there really only remained the ergonomics and modification of the mechanisms to suit the confines of a tank.

There was no screwed-on muzzle cup, as, if a barrel required changing, tank MGs could not be accessed from the outside of the vehicle in action. Instead, a specially designed muzzle booster permitted the barrel to be withdrawn from the rear. It was acknowledged that in practical terms, ranges in excess of 300yd were unlikely to be engaged by the vehicle's machine gun – beyond that the main armament would be employed – but ignoring practicality, the shortened rear sight was still graduated to an optimistic 1,100yd, way beyond the distance at which a tank machine gunner would be able to see the enemy. The guns were equipped with a heavy dovetail mounting underneath the receiver and a pistol-grip trigger was installed midway along the underside of the receiver. This could be unlatched and released if required. A padded face pad similar to that on aircraft guns was attached to the rear of the guns, to enable the gunner to take aim along the top of the gun and prevent injury on rough surfaces.

In other respects, their basic mechanical

functioning was identical to the infantry Mk I. The different marks of AFV guns were broadly mechanically similar, but with minor improvements such as strengthened dovetail bases, improved access for the water hose, left and right feedblocks. Its final incarnation, the Mk VII of 1936, also had a smooth water jacket, earlier models mostly being reworked First World War guns. Virtually all were manufactured at Crayford or converted from Mk I guns at Enfield, but by the time war broke out, it was already becoming apparent that water-cooled rifle-calibre weapons for use in AFVs were deficient in many respects. The biggest problem was exactly the same one that had become apparent in the 1916 tanks: the inability of AFVs to protect the water jackets or carry a condensing canister, which made sustained fire problematical (admittedly, this was not a regular requirement in a tank). In the Second World War, vehicles that had the Vickers were equipped with an 11gal header tank to permit 3,000 rounds of sustained fire if needed. This amount was not arbitrarily decided upon, but was arrived at by the calculation of the quantity of ammunition that could be carried inside the vehicle. It simply took up too much space in an already overcrowded turret and several experiments were undertaken with a pump, or forced cooling via the vehicle's header tank, which eventually became the preferred method.

In the meantime, work had already begun in testing a machine gun of a calibre similar to that of the .50-calibre M2 Browning that had been introduced into US service in 1933. Early American tanks such as the M2 Light Tank and M3 Stuart were well equipped with the powerful Browning, and since the late 1920s Vickers had been investigating the possibilities of producing a scaled-up version of the Mk I gun. In 1933 they introduced the 'Gun, Machine, Vickers 0.5 inch Mk I'. This was a heavily adapted Mk I Vickers,

RIGHT The Cruiser Mk I was also a Vickers tank, designed to move quickly behind enemy lines and cause disruption. Uniquely, it had three Vickers guns, one coaxial with its 2-pounder gun and twin turrets each containing a Vickers gun and gunner. Its crew of six meant it was extremely cramped inside. *(Tank Museum)*

ABOVE Post-1918, the Vickers Company entered into large-scale tank production. This interwar Vickers Medium Mk II was used from 1925 to 1939. Its armament was a 6-pounder QF gun and a .303in Vickers under an armoured shield, leaving little room in the turret for much else. *(Tank Museum)*

BELOW A Carden-Lloyd Tankette, built from 1927 to 1935. These were designed to be reconnaissance and machine gun carriers, able to cross any terrain. As the concept of mobile Vickers guns became less popular in the late 1930s, they were replaced by the Universal Carrier. *(Tank Museum)*

with strengthened components to cope with the larger cartridge, as well as a modified fusee spring and firing mechanism. The .5in Vickers cartridge must not be confused with its American .50-calibre BMG counterpart, as it was 3.18in long and carried a 580-grain cupro-nickel-clad bullet, with a velocity of 2,540fps, whereas the Browning was 3.9in long with a 655-grain bullet that attained 3,030fps. These guns were almost all Mk IIIs with their distinctive MG08-type elongated muzzle boosters, produced for the Royal Navy as anti-aircraft guns; however, the Mk II and the Mks IV and V became the primary armament in light and medium Vickers tanks in the 1930s. On board naval vessels they were normally mounted in pairs or quads and drum-fed by 200-round belts. Vickers claimed that the guns could be reloaded in a mere 30 seconds. They were nicknamed 'Pom-Pom', actually the name given to the earlier Maxim 1-pounder guns, with their slow, measured rate of fire. The quad .5in guns were anything but slow – their 700r/pm rate of fire could launch 50 heavy bullets per second into the air, creating a deadly hail of crossfire for any attacking aircraft to fly through. By the end of the war some 12,500 had been manufactured.

However, the .5in Vickers paled into insignificance compared to the 'High Velocity Class 'D'' guns which were developed in the mid-1920s to meet the growing demand for a heavy anti-aircraft machine gun that could also be mounted on vehicles. This work was undertaken when AFVs were still relatively thin-skinned and a heavy bullet was capable of effective penetration. The guns were classed as 'Gun, Machine, High Velocity 12.7mm', an unusual metric nomenclature for a British-made weapon, but so named to avoid confusion with the existing .5in guns as the 12.7 × 120mm SR calibre was .5in as well, but was not interchangeable, owing to its far longer cartridge case. Two types were made, a 664-grain bullet, designated the V/664, and a later 699-grain V/699, which was to become the standard round. This achieved an impressive velocity of 3,000fps, partly due to the extreme length of the cartridge case (an impressive 4.73in) that gave the bullet a maximum range of 7,000yd, or 15,600ft in an anti-aircraft role.

The guns themselves bore a far greater resemblance to the .303in ground Mk I than did the AFV models, although it was considerably bigger. To increase sales, a double-wheeled infantry version was advertised by Vickers as being man-portable, despite weighing 620lb. Because of the requirement for extracting such a long cartridge case, it had a relatively slow rate of fire of up to 400r/pm, which made it impractical for AA defence when used as a single gun. Despite considerable testing, little could be done to speed up the rate of fire as tests revealed that heat build-up created excessive bore erosion and barrel wear (the barrel alone weighed over 7lb). Mounting as a pair in an AA role proved a far more practical proposition but required an

BELOW Three images of a 12.7mm Class 'D' gun. It has a modified fusee spring housing, crank handle and rear sight, and every component was scaled up to cope with the larger cartridge. A Maxim-type muzzle booster was fitted. *(H. Woodend/NFC)*

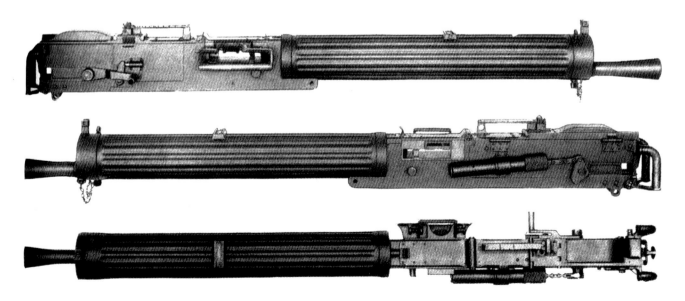

extremely heavy base, frame, gunner's seats, sighting equipment and ammunition mountings and complex traversing gear, all of which pushed the overall weight up to just over a ton. This not only made it necessary to be a fixed installation (or at the very least, vehicle-towed), but also very expensive, and this was probably the main reason for its commercial failure. The British Government did not order any, despite successful testing and only a few were sold to China, Japan and Siam, and few survivors exist today.

There were other important Vickers models produced during the interwar period, such as

ABOVE LEFT A naval quad mount for the .5-calibre Vickers. Note the similarity of the 200-round drums to that of the home-built one pictured being used by the Australian gunner in Amiens in 1918. *(Author)*

ABOVE A 15cwt truck of the Long Range Desert Group unleashes a hail of fire at a distant target from a .5-calibre Vickers. A Lewis gun is wrapped against the elements on a pintle mount just behind the gunner. *(IWM E12408)*

BELOW The .303in Barrel Wear Gauge for the Vickers; it was a go/no go type, calibrated between .303 and .306in. If the gauge could be inserted into the chamber beyond 3.5in, the barrel had to be replaced. *(Richard Fisher)*

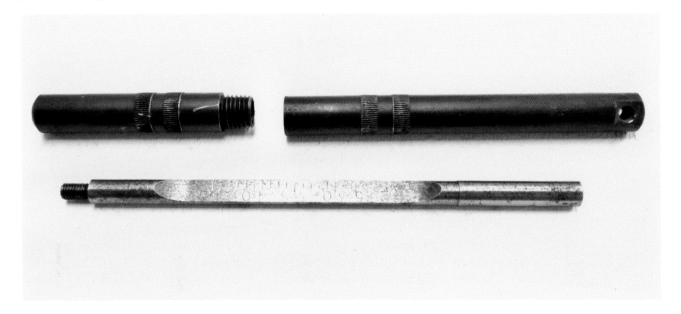

the Bren-like Vickers-Berthier and the gas-operated Class 'K' aircraft guns, but these were not based around the Mk I mechanism, instead relying on air-cooling and box or drum magazines. They were designed to fulfil a different role to that of the rifle-calibre machine guns and so lie outside the scope of this book, but there was one final model of the original Vickers produced that is worthy of inclusion, although it was really nothing more than a Mk I.

The interwar period was a lean time for every armament manufacturer, as public opinion was strongly against military spending. Besides, with the need for reconstruction and the debts incurred during the war, few governments had the money available. Vickers, therefore, had to face the prospect of a reduction in foreign military orders. Neither was there much interest from the British Government; the First World War had cost Britain around £162 billion at today's values. In any case, there were still sufficient Vickers guns in store to last the Army for the foreseeable future. Nevertheless, they needed to continue to manufacture and sell machine guns commercially, so they decided to recommence production of the Class 'C' water-cooled guns.

As befitting peacetime manufacture, these guns were beautifully finished, with deeply blued receivers and green-painted or blued corrugated water jackets. Most were equipped with a new tripod, the Class 'L', which was introduced in 1927. It was of a complex but very efficient design that incorporated a fixed bracket on the right side for an ammunition box as well as having an effective heavy-duty elevating gear and a crosshead mount that could be rotated upwards in seconds to create a high anti-aircraft mount. Every part was produced to blueprint dimensions and individually inspected. The commercial sales of these weapons were primarily responsible for keeping the production lines open for the Vickers water-cooled guns. Large sales were made to Bolivia, China, Portugal and Siam. They were, of course, very costly – almost 50% more expensive than the wartime guns – and today the few survivors represent the pinnacle for collectors of the marque. Not for nothing have these guns and their mounts been described as the 'Rolls-Royce' of machine guns and it is probably not an exaggeration to say that these Class 'C' guns were the finest ever manufactured by Vickers.

The US Model 1915 Vickers

Many countries, including all of the Dominions, adopted the Vickers gun during the First World War, but the only one to actually manufacture their own variant was the United States. In 1912 the US Ordnance decided to test all of the machine guns in service to determine if a single model should be adopted. At the time the military were employing a mix of M1895 Colts, M1909 Benét–Merciés, Lewis light machine guns and M1904 Maxim guns. In extended tests, which included rapid fire and reliability, the Colt overheated severely after 1,000 rounds, leading to cook-offs; the Lewis suffered a bewildering number of stoppages and parts failures; while the Benét jammed if the feed strips were inserted incorrectly and also suffered from frequent firing pin and extractor breakages (for good measure, it also overheated if fired continuously). Only the Maxim worked well, but it was now regarded as too heavy and too costly and the winner by a clear margin was the Vickers. The subsequent report issued by Ordnance Captain J.S. Butler was unequivocal in its praise for the Vickers:

The Board of Ordnance & Fortifications held a meeting on March 15, 1913 to consider the adoption of a new type of machine gun. . . . The Board convened for the competitive test of automatic machine guns at Springfield Armory on September 15, 1913. Seven makes of automatic machine guns were considered and tried out. The Lewis gun during the endurance test had 206 jams and malfunctions, 35 broken parts, 15 parts not broken but requiring replacement as against respectively 23,0,0 replacements for the Vickers gun and 59,7,0 for the Benét–Mercié. The Board is of the opinion that, with the exception of the Vickers gun, none of the other guns submitted showed sufficiently marked superiority for the military service, in comparison with the service Automatic Machine Rifle to warrant further consideration of them in the field test. The Board is of the unanimous opinion that the Vickers Rifle Caliber Gun, Light Model, stood the most satisfactory test. As to the merits of the Vickers gun there is no question – it stood in a class by itself. Not a single part was broken nor replaced. Nor was there a jam worthy of the name during the entire series of tests. A better performance could not be desired.

There was perhaps a little bias existing, as the US Chief of Ordnance had already recommended the Army adopt the Vickers-Maxim 'New Pattern' Model of 1904, but the outbreak of hostilities in 1914 made the supply of Vickers guns impossible as the British economy geared up for full-scale home production. So the order for 4,125 guns was never fulfilled by Vickers, and instead production was to be undertaken by the Colt Company in Hartford, Connecticut, under a contract dated 22 November 1915. This was for an initial 125 guns and 100 tripods, as well as spares kits, belts, boxes, water canisters and tools, the whole being worth a not insignificant $238,472 (£4.7 million today). However, the problem facing Colt was the usual one of production, or rather, lack thereof.

In July 1915 it had agreed to supply Vickers with 6,000 guns which it was patently unable to do. When the balance of the US order for the remaining 4,000 guns was given to Colt, the company was placed in an impossible position. Tooling for manufacturing the guns was still not in place, nor was it likely to be prior to at least spring 1917. Thus Colt reneged on its contract with Vickers, concentrating solely on supplying the US Ordnance Department with its guns and

ABOVE A US M1915 Colt-Vickers showing the modified tripod with its leather securing straps and box-type water container. *(Courtesy Robert G. Segel)*

BELOW LEFT A close-up of the open-pattern muzzle attachment and double claw-type securing clamps for the water hose. These could be easily dislodged, or blow off under pressure, resulting in steam loss. *(Courtesy Robert G. Segel)*

BELOW The most noticeable difference on the Colt-manufactured guns was the machined platform on to which the aperture rear sight fitted. It required considerably more machining than the Vickers type, but the sight was arguably better. *(Courtesy Robert G. Segel)*

ABOVE The front view of the 11mm Colt-Vickers. The air vent holes in the front and the plethora of louvres would have provided more than ample cooling. Only 1,700 of these were manufactured. *(Courtesy International Military Antiques, ima-usa.com)*

spares. By the time America officially entered the war, there were still no Vickers guns in US service, save the original 287 Vickers-Maxim Model 1904s. As the later report on machine gun procurement by Capt J.S. Butler flatly stated, at the start of hostilities with Germany, America had '. . . placed with Colt's . . . an order for 125 Vickers Machine guns, Model of 1915 [and] a further order for 4,000 Vickers machine guns. None of these had, however, been delivered or completed by the manufacturer up to the time of entrance of the United States into the European War, for 6th April 1917.'

This placed the Ordnance Board and the government in an unhappy dilemma, for to cancel the contract would require finding another manufacturer capable of setting up for large-scale production when work was already under way at Colts to produce the required manufacturing tooling. The reality was that Colt must have known that once they were so far down the line it would be impossible for the contract to be withdrawn, and to an extent they had the Ordnance Board over the proverbial barrel. In fairness to Colt, they began to produce the Vickers-Colt Model 1915 guns on 10 May 1917, the first of which was immediately taken for testing, where it fired 40,000 rounds in '. . . a satisfactory manner. There were a number of malfunctions . . . mostly due to a failure to seat the cartridge in the chamber. Four mainsprings, one gib, and one muzzle gland were broken. The Board is of the opinion that the Colt-made Vickers gun is a very efficient weapon.'

Considering that these were the first Vickers ever made by Colt, and they differed from the British-made version in several respects, this was a very creditable performance indeed. Superficially, the guns looked very similar, with fluted jackets, but a closer examination reveals several changes. The tripod design was slightly different; its rear leg was longer and the foot turned upwards. The rear sight lacked the bridge of the British gun, having a 5½in-long leaf-type rear sight with adjustable slider and variable six-hole aperture sight. It was initially graduated to 2,500yd but in 1918 this was altered to 2,800m and it was identical to that of the Model 1917 Browning. The muzzle booster did not incorporate the armoured deflector, being of a more open pattern, and the troublesome threaded attachment for the armoured steam condenser hose was replaced by a quickly detachable double spring-clip, the hoses being made of heavy black rubber, while the spade-grips were not wooden, but of a hard plastic.

However, the major change was invisible – the chambering of the guns for the US .30-06-calibre cartridge. This provided them with a nearly identical performance to the .303in versions, albeit the lighter US bullet (150 grains as opposed to 174) did not provide the extreme range of

BELOW Because the fusee spring needed to be kept properly tensioned, a rod with toothed wheels was attached on the left side of the fusee cover, enabling the pilot to use a hand-wheel to adjust the spring as needed. *(Courtesy International Military Antiques, ima-usa.com)*

the British guns. Reports from the front during early combat indicated that the guns generally performed well; there were some small problems with improperly hardened components, such as recoil plates, lock springs and feed pawls, but these were quality control issues, and not the result of any design failures. Indeed, throughout their service life in the US Army, the Vickers had only nine modifications sanctioned that required changes to be made to factory tooling.

The American divisional machine gun units were trained by British MGC instructors, but their units were allocated in a slightly different manner, every division having 144 guns, but each regiment within it also possessed its own MG unit comprising 12 guns. This meant that an American division had, in theory at least, 192 guns at its disposal, and by the time of the great Allied Offensive of 8 August 1918, there were 13 American divisions equipped with Model 1915 Vickers guns.

If ground guns were plentiful, the same could not be said for air service weapons. Where the US struggled was with the supply of machine guns for their rapidly expanding Air Force, but the work done by Vickers in England on aircraft guns enabled them to copy the most efficient designs. Or they would, had the guns been available. Fortunately the original British contract for 6,000 Russian Vickers guns that had been subcontracted to Colt had been produced at a glacial pace, so when, in March 1917, the Russian Revolution began, 1,200 guns that had been completed but not shipped were held back by Colt and ordered to be modified for air service. This happily coincided with the development by the French of a new 11mm cartridge for aircraft use only, so 802 of the Russian guns were converted. The US Government placed another order in June 1918 for a further 6,000 guns, with 1,700 chambered for the French 11mm cartridge.

The most authoritative report undertaken in 1919 by the Assistant Secretary of State for War, Benedict Crowell, stated that by the end of the hostilities 7,653 American-built Vickers guns were in the hands of the American Expeditionary Force, with a further 3,000 converted for air service. By the end of the war, Colt had produced 12,125 infantry Vickers guns in .30-06 calibre. But it was the end of

LEFT The Vickers-Maxim became a potent symbol during the First World War. This poster exhorting people to buy war bonds dates from 1915. *(Lawrence Brown)*

the line for the Vickers Model 1915 guns, for the United States had by then adopted the Model 1917 Browning, a recoil-operated gun that was simpler to manufacture and maintain than the Vickers as well as being exceptionally reliable. Although it did not arrive in sufficient numbers during the war to replace the Vickers in service (12,780 were produced up to the end of 1918), it rapidly supplanted them post-war, and went on to become America's primary medium machine gun. In a somewhat odd twist, many of the surplus American Vickers were shipped to Britain in 1940 to arm Home Guard units. Their barrels were painted with a red stripe to indicate that they were chambered for .30-06-calibre ammunition.

The ANZAC Vickers

During the First World War, all Dominion and Colonial units serving under British command were required to be issued with the same small arms as the Regular and Territorial armies. Australian forces, who first saw action in the Gallipoli campaign that began in March 1915, formed their own Brigade Divisional Machine Gun units, five in total, which were organised on the same lines as those in the British Army. They had four sections each with

LEFT This Australian Vickers gun has a locally manufactured mount made to enable it to be fired at high angle. The gun crew have also gone to the trouble of making a simple drum to hold the belt, a crude but effective solution to the feed problem associated with AA use. *(Australian War Memorial E 4843)*

four guns comprising a company, and four companies allotted to each of the divisions.

Transporting the guns was, as always, an issue, for while on the Western Front some type of vehicle was generally available, the situation in the Middle East was very different, as distances between the rear and front lines could be dozens of miles, with no forms of transport other than mules, horses, camels or feet available. Probably more than any other Dominion units, the Australians made considerable use of pack animals to carry their guns into action, quite often taking them almost into the trenches to offload. They were greatly aided by the availability of the harness specially devised to carry the Vickers. The teams consisted of a Gun Horse, carrying 160lb, a Tripod Horse carrying 180lb and an Ammunition Horse, carrying 150lb. Other horses ridden by the gun crews led them on a one-to-one basis and these also carried parts, additional ammunition and – at least on one occasion – the company officer's wind-up gramophone. These loads were far beyond anything that could be carried by humans and the animals were worth their weight in gold.

Australian soldiers and their machine gunners in particular earned a reputation for what has been politely described as 'an uncompromising' attitude to fighting Germans, as was evidenced by the fact that no fewer than three Victoria Crosses were earned by their Vickers gunners. They developed a great fondness for the guns

LEFT A pair of Mk I Vickers in the care of No 3 Section, 2nd New Zealand Machine Gun Corps, at Homs, Syria, 1918. Despite the desert heat, the guns coped admirably, using only slightly more water than usual, but internal parts had to be sparsely lubricated as the dust turned oil into grinding paste. *(Australian War Memorial B00381)*

ABOVE No 2 platoon, 14th Brigade New Zealand Machine Gun Company, Vella Lavella Island, South Pacific, with their Vickers gun. This is an early First World War 'five-arch' model, with dial sight. The motley crowd behind the gun are armed with an eclectic mix of Short, Magazine Lee-Enfields, M1928-A1 Thompsons and a couple of M1 .30-calibre carbines. *(Wellington War History Collection, Alexander Turnbull Library. WH 241)*

and most of those issued to Australian Imperial Forces during the war found their way back to Australia at the end of hostilities. These were employed for training but many were well past their prime by this time. The quantities of spares the Australians possessed, though considerable, would not last for ever if the guns were to continue in military service. So, in 1929, it was agreed that Vickers guns would be produced at the government-owned facility at Lithgow near Sydney, which had opened in June 1912 to provide Lee-Enfield service rifles. Post-war, against the current thinking of most of Europe, the Australian Government decided not to reduce, but to expand the factory site, so between 1925 and 1930 it was extended to make provision for a Vickers production facility.

Using American Pratt & Whitney tooling it became the first fully electric factory in the country. Admittedly, the demand for the guns at this time was minimal and production initially centred around spares and barrels, but in 1929 the first Mk I Vickers rolled off the line, in the face of strong criticism from the press, public and many politicians who believed, understandably, that the money could be better spent.

ABOVE Australian machine gunners march out of the fighting at Contalmaison, Somme, July 1916. The front gun has its Sangster tripod in place, and the men following are carrying canvas water containers, spares boxes and other vital items. *(Australian War Memorial PO 3137I)*

RIGHT Both Maxims and Vickers guns saw service at Gallipoli. This gun of the 6th Australian Light Horse Regiment was one of the first to have an impromptu anti-aircraft mount made and it wears a leather jacket cover. *(Australian War Memorial AO2355)*

The wisdom of the government's decision was to be shown in 1938 as another war loomed, when Lithgow became the only factory able to produce guns outside the United Kingdom. Additional machinery was installed and production of the Mk IV tripod and guns was increased, supplied by 'feeder' factories located near to the main manufacturing facility; the total number of workers exceeded 12,000 by 1943. By the time production ceased in June 1945, 10,170 Mk I Vickers guns with tripods had been made, as well as thousands of barrels and tons of spare parts.

These guns varied only slightly from the British-manufactured ones, all having a plain steel water jacket, and improvements were made to some components, in some instances with the introduction of higher-quality metals to enable such items as feedblocks and locks to be made lighter but stronger (although all parts remained interchangeable with any other Vickers). Unlike the facilities at Erith and Crayford, Lithgow also had the capacity to make variants such as the Mk XXI Vickers, a

RIGHT Tripods at the Lithgow factory undergoing final inspection before being painted. *(Australian War Memorial 001746)*

CENTRE Jackets on Lithgow Vickers guns being rubbed down to remove burrs. *(Australian War Memorial 001765)*

simplified pistol-gripped version for AFVs, and the air-cooled Mk V for aircraft. Most of these guns saw hard combat across the Western Desert and in the Pacific in the hands of seven Machine Gun Battalions (numbered from the 2/1st to the 2/5th and then the 6th and 7th) who were trained on exactly the same lines as their fathers had been between 1915 and 1918.

The guns were also used to great effect to provide long-range supporting fire during the Korean War. In 1958 they too were replaced by the FN MAG and stored, but unusually, in the early 1980s, the decision was made to sell off the remaining guns and spares into the collectors' market, and for a decade or so, Australian Vickers guns could be purchased complete for around £450. This accounts for the relatively large number of Lithgow Vickers still in the hands of collectors in Britain and the USA, as well as some earlier British-manufactured guns that had been in reserve stores alongside them.

New Zealand's machine gunners also played a major part in the First World War. In 1896 their Army had purchased six Maxim-Nordenfelt guns, increasing the number to 35 by 1901. The biggest problem facing them was in replacing damaged or worn parts as no manufacturing facilities existed in New Zealand. That being the case, everything had to be carried by sea from the UK, and from order to delivery could take up to six months. The situation changed dramatically after Gallipoli, when the New Zealand Expeditionary Force (NZEF) was sent to Egypt from where, in January 1916, it raised three companies of the New Zealand Machine Gun Corps (NZMGC), each of which was attached to their respectively numbered brigades. By 1918, this had been increased to five companies. They were equipped with British-manufactured Vickers guns and tripods and trained in exactly the

BELOW A 1944 Lithgow-manufactured Vickers. Aside from the smooth jacket and armoured nose cone, it is identical externally to a First World War gun. To speed up production, there were some internal modifications, but all parts were still interchangeable with British-made guns. *(Ian Durrant)*

same manner as British MG units. They fought on the Western Front and against the Turks in Palestine, among whom their guns gained great respect for their deadly efficiency. Indeed, the capture of two key towns El Arish and Magdaba in December 1916 was stated by the official historian to have been as a result of '. . . the fearsome level of overhead fire laid down by the MG units'.

After the war, the NZMGC units also returned home with their Vickers guns, but much training during the inter-war period and a lack of spares meant that by the commencement of hostilities in 1939 the NZMGC could muster only 12 functioning guns per company, instead of the officially required 16 and many of these were worn out. When they began training late in 1939, these were supplemented by brand-new Lithgow-made guns, to the delight of the crews, who accepted with good grace the slow and messy business of removing several pounds of solidified Cosmoline grease from each one.

The New Zealanders fought across Greece, Italy and North Africa, the 27th MG Battalion recording that over the Italian campaign their guns alone fired almost 9 million rounds. In the mountain terrain, they faced the same transportation problems as in the previous war, outlined by one company commander: 'Whilst two men could carry the gun well forward into action, the [ammunition] portage was another question. For ample supplies of ammunition it [the MG unit] had to rely on getting its transport well forward and this transport had to compete with other much-needed stores.' Put into perspective, such a quantity of ammunition required 36,000 ammunition boxes to be carried over a period of time into the firing positions.

If that appeared difficult, it paled into insignificance when both the Australian and New Zealand machine gunners found themselves fighting in the jungles of New Guinea and the South Pacific, where thick vegetation, virtually no lines of supply and energy-sapping humidity hampered operations. This provided the two MG companies allotted to the 8th and 14th Infantry Brigades with their own unique problems. One of the biggest was the continual humidity penetrating the ammunition tins as soon as they were open, making the belts soggy and forming verdigris on the brass, leading to constant stoppages. There was no easy cure for this, and MG teams were told to keep ammunition sealed until the last possible minute, while barrels and internal parts had to be oiled daily. No Vickers were employed by NZ troops in Korea, and in 1965 they were officially replaced by the almost-universally adopted FN MAG general-purpose machine gun.

Canadian Vickers

Unlike any other British Dominion, Canada's officially adopted machine gun was the Colt Model 1914 (a development of the Model 1895 'potato digger') and the first Canadian Expeditionary Force soldiers of the 2nd Division who arrived in France in December 1914 were equipped with them. The problem with the Colt was that it suffered from a number of deficiencies that only came to light when it began to be employed in tasks for which it had not been designed, such as providing heavy overhead fire. The mechanism had an external gas lever that did not permit an elevated firing position to be accurately maintained, leading to rounds dropping short – dangerous for friendly troops under the arc of fire. Nor were the tripods as stable as the Mk IV and the joints suffered premature wear, often requiring wooden wedges to keep the barrels correctly

BELOW The only machine guns in Dominion service in 1914 apart from the Maxims were the Colt M1895 'potato diggers' initially used by the Canadian Expeditionary Force. The U-shaped reciprocating arm that gave it its nickname can be seen under the barrel. Its high profile and inability to sustain high rates of fire made it impractical.
(Ian Durrant)

LEFT Two MMGS sidecar outfits, almost certainly Clynos, block a road junction during manoeuvres in England. The enormous loads they had to carry can be seen here; each motorcycle was expected to haul around 600lb of men, guns and equipment. *(Author)*

aligned. In fact, both guns and mounts suffered from accelerated wear due to the muddy conditions in which they fought. Neither did the British Army, under whose command the Canadians fought, have access to spare parts, so it was deemed entirely sensible for the Vickers – a far superior weapon – to be issued to replace them once sufficient Mk I guns were available. As 300 guns were needed immediately this did not actually occur until July 1916, when enough Vickers guns had been made to permit an allocation to the Canadians.

The four Canadian infantry divisions were, unsurprisingly, initially organised in the same manner as their British counterparts. Each was allocated two MG companies, each of four platoons of eight guns, which gave them a field strength of 64 guns per battalion. Unlike the British, who had formed the MGC in 1915, until the spring of 1917 they remained attached as MG battalions to their parent brigades, although another reorganisation a year later turned these brigade companies into numbered machine gun battalions. They also had two motor machine

BELOW Two unidentified British soldiers in a light car equipped with a Vickers gun, probably photographed in Palestine, 1917. The steel cradle holding the ammunition box is visible on the side of the gun. *(Australian War Museum P08997-03)*

RIGHT An iconic image from the Battle of Passchendaele. Reginald Le Brun of the 16th Canadian MGC is looking directly at the camera from behind his Vickers gun, with his doomed crew behind him. *(Canadian War Museum, 19930013-509)*

BELOW Canadian machine gunners cleaning an armoured Autocar. These 2½-ton vehicles carried two Vickers guns, one Lewis gun and a crew of eight. The Vickers guns were pintle-mounted and could be fired from either side of the vehicle. *(Canadian War Museum 3395367)*

gun brigades, one formed in 1916, the second in 1918, the Clyno motorcycles of the original battalion by then having been replaced by far more practical open-backed trucks.

In combat, the Canadians had learned a great deal about Vickers tactics from early experience in Flanders and on the Somme, and they were less restricted by traditional military thinking where the new weapon was concerned.

They were more flexible and always prepared to try new tactics, which probably reached their apogee during the very successful Vimy Ridge campaign where 294 Vickers guns were employed to provide a creeping MG barrage in conjunction with the preliminary artillery barrage, effectively boxing in the German defenders and preventing their reinforcements from reaching the front lines – the first time this tactic was

tried. The statistics for organising this feat were impressive indeed: 20,000 boxes of ammunition had to be brought up to the battery positions, as well as 600 gun barrels, 1,470 gallons of water and half a ton of spares.

When they opened fire at 5:30am on 9 April, the guns' overhead barrage was so effective that the subsequent report on the battle specifically mentioned the part played by the Vickers companies. As a measure of reward and acknowledgement, on 16 April 1917 they were re-formed as the Canadian Machine Gun Corps (CMGC). They played a further major role in the advances of the final Hundred Days Offensive, which began on 8 August 1918, with Vickers teams moving up with the infantry to provide heavy covering fire on the entrenched German rearguards, enabling their own troops to advance successfully.

At the end of hostilities, the Canadians (as did their Australian and New Zealand counterparts) triumphantly returned home with their guns, estimated to be 488 in total. These suffered from the same over-use as all the others and when the Second World War was declared, they were allocated to 14 new machine gun regiments comprising one battalion each. Alas, between them, the regiments could muster only 192 working guns, which was clearly insufficient, so armouries, drill halls and even regimental museums were scoured and a further 112 that had been designated for drill or exhibition purposes and were marked 'DP' (denoting drill purposes) were subsequently refurbished. As in New Zealand, there were no manufacturing facilities in Canada for production and the only solution was to issue new guns to Canadian units as they arrived in the UK.

ABOVE A Mk I Canadian-manned Universal Carrier with its Vickers gun on its tripod. The gun has the angled metal carrier fitted for the ammunition box, designed to enable it to feed without assistance from the team's No 2. The gun transit case has been bolted to the front of the vehicle. *(Tank Museum)*

LEFT Artificers at the Royal Small Arms Factory, Enfield, refurbishing old Vickers guns prior to their reissue in the Second World War. *(Author)*

BELOW A drawing from a Russian magazine of 1911 showing early attempts at producing a machine gun, and the 'new' Russian Model 1910 Maxim. Russians were subsequently taught that it had been invented, not in England, but in Russia.
(Peter Smithurst)

They were to see hard service during the defence of Hong Kong and the Dieppe Raid, both of which resulted in the loss of many precious guns. Vickers also came ashore with Canadian MG units during the D-Day landings on 6 June 1944 at Juno Beach, where they were able to lay down heavy suppressing fire on some of the dozens of German concrete MG emplacements, enabling troops to advance. Having fought across northern Europe, in 1945 they were returned to the Central Ordnance depot in Montreal, for the inevitable storage, some being very briefly taken to Korea. By then, the Canadian War Office had followed the line of current thinking, believing that they were outdated and in 1958 they were declared surplus to requirements and struck off strength. A small number were offered to museums and sold into the collectors' market, but most were taken apart and sold off as spares to countries that still retained the Vickers in service, such as India, Pakistan and a large number to Egypt. Canada opted to have a variant of the Browning machine gun as its service medium machine gun until the 1980s.

The Russian Vickers-Maxims

Of all of the countries to adopt the Vickers or Maxim (in excess of 30, not including Dominions) the one power that made the single greatest use of the guns was Russia, and they alone developed and manufactured their own variant, the *Pulemyot Maxima obraztsa 1910 goda* or Maxim's Machine Gun, Model of 1910. Initially, they had bought just two Light Pattern guns in 1909 and from the outset they had planned to manufacture their own copies. However, when the country was plunged into

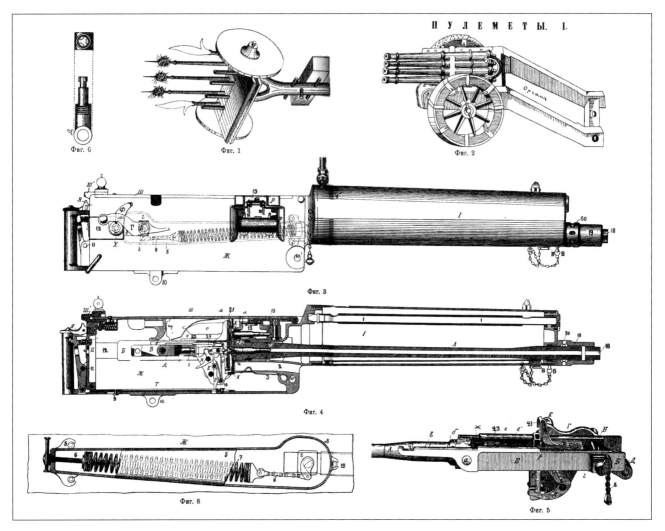

war in 1914, the need for working guns became paramount, more so after the disastrous early battles of Tannenberg and Komarów. Russia was by then desperate for machine guns but despite setting up a production facility at the Tula Arsenal in Tula Oblast, 120 miles south of Moscow, the very slow setting-up time for beginning the complex manufacturing operations meant that in the interim they had to look elsewhere.

So on 12 October 1915, an order for 10,000 guns was placed via Vickers with Colt in Hartford, but as we have seen this merely served to compound Colt's existing manufacturing woes, as they were totally incapable of meeting such a colossal order. By the time of the October Revolution in 1917, it is believed that at most, some 3,000 guns had been supplied. However, during this time the Russians had expanded their existing production facilities and begun to make copies of the Vickers New Light Model of 1906 in the guise of the M1910. These were essentially a clone, but with steel instead of brass to save weight and cost, the M1910 being 10–12lb lighter than the Vickers-Maxim, which was necessary as its wheeled carriage alone weighed 80lb, giving a total weight of around 145lb. It differed little mechanically, incorporating the simplified lock and modified cocking handle of the Vickers design but was, of course, chambered for the Russian 7.62 × 54mm R cartridge with a 148-grain bullet.

In 1930 a new cartridge was produced, the M1930 Type D with a 182-grain boat-tailed bullet, providing similar performance to the British Mk VIIIz ammunition, which enabled the M1910 to fire in excess of 4,500yd. Probably the most distinctive change, and one that sets the Russian guns apart from all other Vickers variants, was the method of filling the water jacket. The small filler orifice was hopeless in freezing conditions, so the Russians copied captured Finnish commercial Maxims (designated the M09/21) that had been modified with the addition of a large hinged filler cap on the top of the jacket, nicknamed the 'tractor cap'. This enabled not only water to be quickly poured in, but also snow or ice. By 1939 all Soviet M1910 guns were being manufactured with this filler in place. By the end of the Second World War, the Russians too had realised that the weight of the M1910s and their requirement for water-cooling could create serious tactical limitations. After facing devastating fire from the thousands of German MG34 and MG42 light machine guns, the Russians had, by 1945, replaced most of their M1910 guns with much lighter and simpler air-cooled guns, such as the *Pulemyot Degtyaryova Pekhotny* or the Degtyaryov DP-27, of which nearly a million were manufactured. The huge numbers made mean that Russian M1910s still turn up around the world, several having been captured recently in Afghanistan.

BELOW An early production M1910 on the Sokolov mount. It has the normal small water-filler orifice just in front of the receiver. *(Courtesy Trustees of Royal Armouries Museum)*

BELOW A Second World War-manufactured M1910 with the 'tractor cap' filler on the water jacket. Although the weight of the shield and mount made it considerably heavier than the Vickers, it was far more portable in combat. *(Courtesy Trustees of Royal Armouries Museum)*

Chapter Eight

Maintaining and shooting a Vickers

A Vickers gun, with its myriad of parts and potential for mechanical failure, was a complex weapon to set up and maintain. How this was done and the sequences for stripping and cleaning a gun are examined along with detailed supporting images and technical drawings covering every part of the guns and their components.

OPPOSITE A spring gauge being used to measure the weight of the pull of the fusee spring on the crank handle. Its pull-weight is calibrated from 0 to 20lb. Adjusting the fusee spring required the T-shaped bar at the forward end of the fusee cover to be tightened or loosened. *(Author)*

Plate A₁

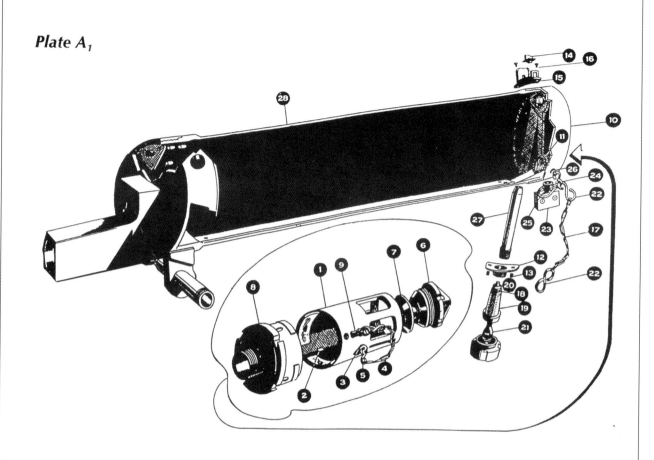

Ref. No.	Designation	Vocab. Number	Mat.	No. Off	Drawing Number	Remarks
	Plate A₂ **GUN, MACHINE, VICKERS, .303-in. MK 1** *(continued)*					
...	CASING, BARREL *(continued)*	...	...	...	...	Mk I is corrugated, Mk I* smooth
29	TROUGH	BD 0636	Brass	1	MGD 212	BD 8095 in UK lists
30	RIVET, front	CA 0443	Steel	2	MGD 157	UK Vocab. No., no Aust. listing
31	RIVET, rear	CA 0456	Steel	1	MGD 158	UK Vocab. No., no Aust. listing
...	TUBE, STEAM, No. 1 #	BD 0985 SA	...			
32	ACORN	CA 0526	Brass	1	MGD 1	UK Vocab. No., no Aust. listing
33	RIVET	CA 0529	Brass	1	MGD 153	UK Vocab. No., no Aust. listing
34	CYLINDER	CA 0527	Brass	1	MGD 40	UK Vocab. No., no Aust. listing
35	HEAD, screwed	CA 0528	Brass	1	MGD 55	UK Vocab. No., no Aust. listing
36	SCREW, keeper, Mk I #	BD 0908	Steel	1	MGD 172	
37	SLIDE-VALVE	CA 0530	Brass	1	MGD 183	UK Vocab. No., no Aust. listing
...	CASING, BREECH, MK I	A	...		MGA 226	
38	BLOCK, trunnion, Mk I	...	Steel	1	MGD 13	
39	BUSH, distance	CA 0457	Steel	1	MGD 42	UK Vocab. No., no Aust. listing
40	PLUG, screwed, No. 1 #	BD 0873	...	1 ‡	MGD 127	
41	HEAD	BD 0735	Fibre	1 ‡	MGD 54	Composition
42	RIVET	...	Brass	2 ‡	MGD 146	
43	WASHER	...	Brass	4 ‡	MGD 219	
44	LINK	...	Steel	1 ‡	MGD 69A	
45	LOOP	...	Steel	1 ‡	MGD 69	
46	STUD	BD 0972	Steel	1 ‡	MGD 208	
47	SOCKET, acorn, steam tube, No. 1 #	BD 0972	Steel	1	MGD 184	
48	RIVET	CA 0460	Steel	3	MGD 154	UK Vocab. No., no Aust. listing
49	ADAPTER, condenser, Vickers, .303-in. M.G., Mk I #	BD 0005	...	1	...	

\# Component part provided for normal maintenance; held in ordnance store for issue.
‡ Parts marked thus are used elsewhere on the gun; the 'No. off' applies only to each particular Ref. No.

The Vickers is something of a mechanical anomaly, employing a simple mechanical process, namely a gas-assisted short recoil system that operated a cyclical loading sequence, which was at the same time an extremely complicated concept to turn into a mechanical reality. Leaving aside the requirement for learning to aim, direct and fire at long range, merely setting up a Vickers gun in order to fire it required considerable training and an intimate knowledge of every functioning part. Prior to a gun being assembled on its tripod for firing, a series of safety checks must be undertaken to ensure functionality. We shall assume that this work was being carried out on a gun that was not in action, is unloaded and cold. (If the gun had been firing, the ammunition source must first be removed, so the belt is taken out of the feedblock and the block removed. The crank handle is worked twice, the trigger pulled and the breech visually inspected.)

Firstly, the water jacket should be emptied via the front drain plug by removing the condenser hose. Then an external visual check was needed to look for obvious problems; one in particular that dogged all water-cooled machine guns was penetration of the water jacket by projectiles. Tiny holes could be ignored *pro tem* but larger ones could not, and gunners used the patches supplied in the spares kit to make repairs. It was a temporary fix, adequate until such time as an armourer was able to soft-solder a metal patch in place. If the gun looked outwardly sound, a detailed inspection of the internals came next. Cleaning or replacing the barrel would be the first task. Under normal usage this would be required after 10–12,000 rounds assuming a moderate rate of fire of 200r/pm had been adhered to over a period of about an hour – a recommended rate of expenditure that enabled the barrel to stay relatively cool. However, firing at the gun's maximum suggested rate of 500r/pm would increase wear dramatically, requiring the barrel to be changed after 2–3,000 rounds. The forward top cover is opened, and the feedblock removed. The rear top cover is then

ABOVE The ammunition source must first be removed; the bar underneath the feedblock releases the belt from the pawls, enabling it to be pulled free. *(Author)*

BELOW LEFT The gun is cocked twice to eject any live rounds, the feedblock is then removed and the chamber checked. *(Author)*

BELOW The fusee cover is unlatched and the spring unhooked from the fusee chain. *(Author)*

Plate A₂

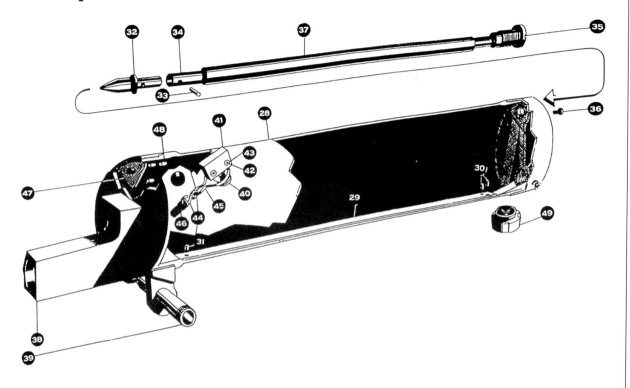

Ref. No.	Designation	Vocab. Number	Mat.	No. Off	Drawing Number	Remarks
	Plate A₁ GUN, MACHINE, VICKERS, .303-in. MK 1	BD 7930 GA			MGA 790	BD 0600 GA fitted with the dial sight bracket.
...	CASING, BARREL	A				
...	ATTACHMENT, MUZZLE, BALL, MK I #	BD 0768 SA	...			
A 1	CASING, outer, Mk I #	BD 0636	Steel	1	MGD 22	
2	PIN, stop	...	Steel	1	MGD 112	
3	STUD, screwed plug #	BD 0972	Steel	1 ‡	MGD 208	
4	CHAIN, K.G., (8 links)	BD 0195	Brass	1 ‡	MGD 29	
5	S-HOOK, M.G. #	BD 2027	Steel	2 ‡	MGD 167	
6	CONE, front, Mk III #	BD 0647	Steel	1	MGD 32	
7	DISC #	BD 0680	Steel	1	MGD 41	Thin, sheet steel pressing
8	GLAND, No. 1, Mk I #	BD 0715	Steel	1	MGD 48	Packing is part no. BD 0716
9	PIN, split #	BD 0836	Steel	1	MGD 109	Machined split pin
10	CAP, end	BD 7690	Steel	1	MGD 18	Mk I. Soldered on
11	BARREL-BEARING, front, with muzzle guide	...	Brass	1	MGD 44	
12	BOSS, condensor	...	Steel	1	MGD 33	
13	SCREW, securing	BD 8015	Steel	2	MGD 173	or rivetted
14	FORESIGHT, Mk I #	BD 0924	Steel	1	MGD 179	Blade
15	BRACKET	BD 7669	Steel	1	MGD 11	RH wing milled; LH, punch hole
16	SCREW, securing	BD 8014	Steel	1	MGD 174	
...	PLUG, cork #	BD 0871 SA	...			
17	CHAIN, M.G., (11 links)	...	Brass	1 ‡	MGD 26	
18	COLLAR	...	Brass	1	MGD 30	
19	CORK #	BD 0650	Cork	1	MGD 34	
20	PIN, fixing #	BD 0815	Brass	1	MGD 93	
21	STEM	...	Brass	1	MGD 200	
22	S-HOOK, M.G. #	BD 2027	Steel	2 ‡	MGD 167	
23	PLUG, screwed, No. 1 #	BD 0873	...	1 ‡		
24	LINK	...	Steel	1 ‡	MGD 69A	
25	LOOP	...	Steel	1 ‡	MGD 69	
26	STUD #	BD 0972	Steel	1 ‡	MGD 208	
27	TUBE, steam escape	...	Steel	1	MGD 214	
28	CASING, JACKET MK I*	...	Steel	1	AID 1704	Corrugated is Mk I, early model. Jackets hold 10 pints of fluid

\# Component part provided for normal maintenance; held in ordnance store for issue.
‡ Parts marked thus are used elsewhere on the gun; the 'No. off' applies only to each particular Ref. No.

lifted, the crank handle is pulled back and the lock pulled upwards and removed by being given a third of a turn in either direction to free it from the conrod. This was then placed on a clean cloth for later inspection.

The fusee cover is unlatched and the spring disconnected. The muzzle booster is unlocked and removed, enabling the muzzle cup to be unscrewed, often a difficult task involving a heavy mallet and C spanner. Sometimes well-used cups could be virtually heat-welded in place on the muzzle and would require so much force to remove that they became scrap anyway, but until it was unscrewed the barrel could not be removed. Experienced gunners would regularly oil the cup thread to aid future removal. The barrel in a Vickers gun has two sets of sealing glands to prevent loss of coolant water. The sealing material is asbestos string, liberally coated in oil, wound around the machined grooves at the chamber end of the barrel and at the muzzle.

There is something of an art to getting this correct; too much packing will hamper the recoil action of the barrel, slowing the rate of fire and causing stoppages. Enough must be wound round to seal the gap and prevent water loss while still allowing for barrel expansion when hot. Machine gunners generally found that a length of string twice the circumference of the jacket was fine for the front gland, and one and a half times the length of the water jacket was sufficient for the rear cannelure. The first sign that a gland is incorrectly packed is water leaking from the breech or muzzle. The suggested replacement time for the packing was after 5,000 rounds had been fired.

Once the new barrel has been inserted, the replacement muzzle cup must be screwed on and preferably this should be a new one. The

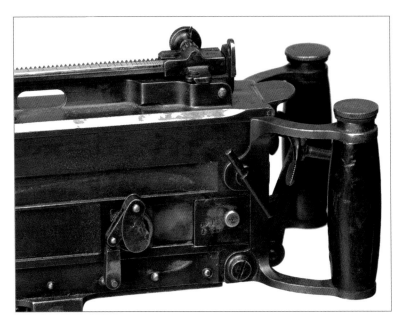

ABOVE A close-up of the fusee and its chain with the spring removed. The 'T' pin securing the handgrips can be seen at the rear of the receiver. *(Author)*

BELOW With the top cover opened, the lock is removed. *(Author)*

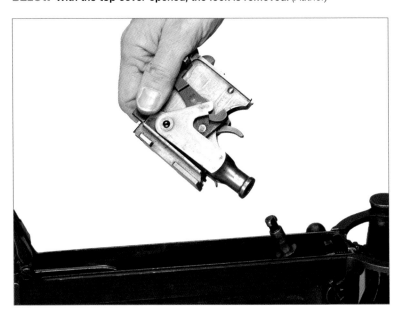

LEFT A close-up of the connecting rod and its locking joint. *(Author)*

Plate B

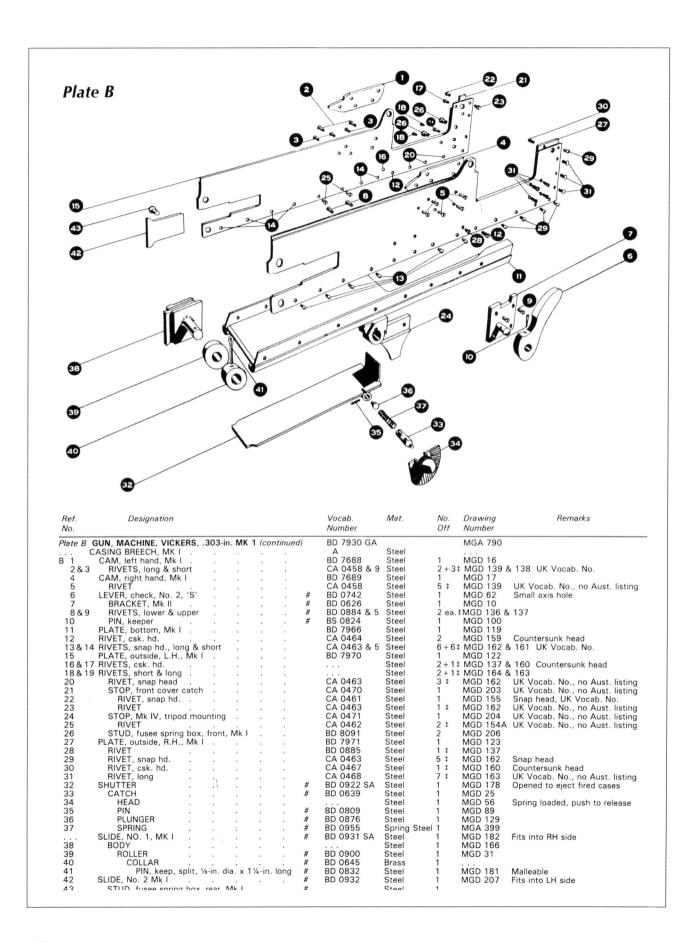

Ref. No.	Designation	Vocab. Number	Mat.	No. Off	Drawing Number	Remarks
Plate B	**GUN, MACHINE, VICKERS, .303-in. MK 1** (continued)	BD 7930 GA			MGA 790	
. . .	CASING BREECH, MK I	A	Steel		. . .	
B 1	CAM, left hand, Mk I	BD 7688	Steel	1	MGD 16	
2 & 3	RIVETS, long & short	CA 0458 & 9	Steel	2 + 3‡	MGD 139 & 138	UK Vocab. No.
4	CAM, right hand, Mk I	BD 7689	Steel	1	MGD 17	
5	RIVET	CA 0458	Steel	5 ‡	MGD 139	UK Vocab. No., no Aust. listing
6	LEVER, check, No. 2, 'S' #	BD 0742	Steel	1	MGD 62	Small axis hole
7	BRACKET, Mk II #	BD 0626	Steel	1	MGD 10	
8 & 9	RIVETS, lower & upper #	BD 0884 & 5	Steel	2 ea.‡	MGD 136 & 137	
10	PIN, keeper #	BS 0824	Steel	1	MGD 100	
11	PLATE, bottom, Mk I	BD 7966	Steel	1	MGD 119	
12	RIVET, csk. hd.	CA 0464	Steel	2	MGD 159	Countersunk head
13 & 14	RIVETS, snap hd., long & short	CA 0463 & 5	Steel	6 + 6‡	MGD 162 & 161	UK Vocab. No.
15	PLATE, outside, L.H., Mk I	BD 7970	Steel	1	MGD 122	
16 & 17	RIVETS, csk. hd.	. . .	Steel	2 + 1‡	MGD 137 & 160	Countersunk head
18 & 19	RIVETS, short & long	. . .	Steel	2 + 1‡	MGD 164 & 163	
20	RIVET, snap head	CA 0463	Steel	3 ‡	MGD 162	UK Vocab. No., no Aust. listing
21	STOP, front cover catch	CA 0470	Steel	1	MGD 203	UK Vocab. No., no Aust. listing
22	RIVET, snap hd.	CA 0461	Steel	1	MGD 155	Snap head, UK Vocab. No.
23	RIVET	CA 0463	Steel	1 ‡	MGD 162	UK Vocab. No., no Aust. listing
24	STOP, Mk IV, tripod mounting	CA 0471	Steel	1	MGD 204	UK Vocab. No., no Aust. listing
25	RIVET	CA 0462	Steel	2 ‡	MGD 154A	UK Vocab. No., no Aust. listing
26	STUD, fusee spring box, front, Mk I	BD 8091	Steel	2	MGD 206	
27	PLATE, outside, R.H., Mk I	BD 7971	Steel	1	MGD 123	
28	RIVET	BD 0885	Steel	1 ‡	MGD 137	
29	RIVET, snap hd.	CA 0463	Steel	5 ‡	MGD 162	Snap head
30	RIVET, csk. hd.	CA 0467	Steel	1 ‡	MGD 160	Countersunk head
31	RIVET, long	CA 0468	Steel	7 ‡	MGD 163	UK Vocab. No., no Aust. listing
32	SHUTTER #	BD 0922 SA	Steel	1	MGD 178	Opened to eject fired cases
33	CATCH #	BD 0639	Steel	1	MGD 25	
34	HEAD	. . .	Steel	1	MGD 56	Spring loaded, push to release
35	PIN #	BD 0809	Steel	1	MGD 89	
36	PLUNGER #	BD 0876	Steel	1	MGD 129	
37	SPRING #	BD 0955	Spring Steel	1	MGA 399	
. . .	SLIDE, NO. 1, MK I #	BD 0931 SA	Steel	1	MGD 182	Fits into RH side
38	BODY	. . .	Steel	1	MGD 166	
39	ROLLER #	BD 0900	Steel	1	MGD 31	
40	COLLAR #	BD 0645	Brass	1	. . .	
41	PIN, keep, split, ⅛-in. dia. x 1¼-in. long #	BD 0832	Steel	1	MGD 181	Malleable
42	SLIDE, No. 2 Mk I	BD 0932	Steel	1	MGD 207	Fits into LH side
43	STUD, fusee spring box, rear, Mk I #		Steel	1		

muzzle booster is then locked into place with a half-turn and secured by a split pin. The water jacket can be filled (taking around 7pt) via the rear filler orifice, but the jacket screw plug must be tightened properly, otherwise steam will escape. It is also imperative that the front plug is in place at this point, or the water will be lost.

The lock must then be examined for broken springs or a damaged firing pin. It is extremely easy to disassemble if these require replacement, requiring one pin to be pressed out. A spare lock, two feedblocks and all internal parts were carried in the spares box. Dummy cartridges should be put into the carrier to ensure all is working smoothly, and live ammunition should *never* be used for this, as the lock can be hand-cocked and the trigger released by finger pressure only, resulting in the cartridge being fired. Firing-pin protrusion needs to be checked; somewhere between .058in and .065in is ideal. The gunner should measure the pull-weight of the lock spring with the spring balance. It should begin to release the tumbler at around 15lb. All lock parts must move freely, and the carrier rise and fall to its correct height. The machined grooves on its face that grip the cartridge should be closely examined for cracks or breaks, not uncommon if the metal quality or hardening are poor and the flanges on the base of the lock can sometimes shear. Once the gunner is satisfied, the lock is lubricated and refitted on the conrod. As a new barrel has been fitted, the cartridge headspacing will be required, as the new barrel will be of fractionally different dimensions. This will result either in a gap between the cartridge base and lock-face, or such a tight fit between the two that the lock will jam when closed. If the headspace is wrong and the gun fired, damage may occur to the barrel trunnion block, rear of the barrel or the face of the lock. The lock was designed by Vickers engineers to enable this to be done easily by inserting a dummy cartridge into the chamber. Having disconnected the fusee spring, the lock is pushed home gently using the crank handle. If it stops before reaching the check lever on the side of the receiver, or slides fully home without hindrance, then the headspace must be corrected. A series of thin washers, No 1 (.003in) and No 2 (.005in) can be used in combination to achieve the perfect setting.

The cup is then unscrewed, assuming it has not been welded into place by fouling, in which case a 'C' spanner and hammer will be required. *(Author)*

When correctly set up, the lock should resist slightly before the crank handle goes fully home.

The next check was on the fusee spring itself; this is as vital a component as the barrel or lock, and easy to overlook. Its importance was drummed home to the gunners, for failure or incorrect setting up could cause the gun to fire erratically, very slowly or not at all. It is attached to the fusee chain which enables the spring to wind itself up. This spring is very strong indeed, and it was recommended in the Vickers manual that it be detached prior to any work being carried out on the lock or internal parts. At this point the internal sideplates must be examined, as the springs on them grip the carrier, preventing it from dropping down when the lock is pulled back. This would result in it being in the wrong position when the lock is pushed forward to cock the gun, so there would be a subsequent failure to collect a fresh cartridge from the belt. The fusee spring can then be reconnected and must be gauge tensioned to between 7 and 9lb, and as the gun fires, this must be maintained by turning the screw adjuster; one and a half turns being equal to a ½lb increase in tension but loosening the spring to less than its recommended tension will cause the gun to fire faster, which if uncorrected would eventually result in damage or parts failure.

At this point, the timing of the gun should be checked to ensure the firing pin releases exactly at the moment it is in contact with the cartridge primer. This is done with both top covers open,

Plate C

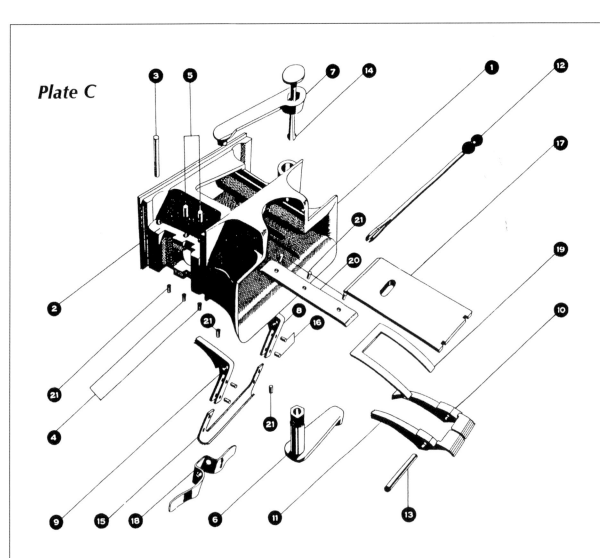

Ref. No.	Designation	Vocab. Number	Mat.	No. Off	Drawing Number	Remarks
	Plate C GUN, MACHINE, VICKERS, .303-in. MK 1 *(continued)*	BD 7930 GA	...		MGA 790	
...	BLOCK, FEED, RIGHT HAND, MK I #	BD 0620 A	...	1	MGA 224	All Mk I guns are RH feed
C 1	BODY #	BD 7653	Steel or Gun Metal	1	MGD 6	Also made from steel in UK
2	GUIDE, cartridge & stop	CA 0440	Steel	1	MGD 51	UK Vocab. No., no Aust. listing
3	RIVET, long	CA 0442	Steel	1	MGD 142	UK Vocab. No., no Aust. listing
4	RIVET, bottom	CA 0443	Steel	2 ‡	MGD 157	UK Vocab. No., no Aust. listing
5	RIVET, top	CA 0443	Steel	2 ‡	MGD 157	UK Vocab. No., no Aust. listing
6	LEVER, bottom #	BD 0740	Steel	1	MGD 61	
7	LEVER, top #	BD 0753	Steel	1	MGD 67	
8	PAWL, bottom front #	BD 0795	Steel	1	MGD 77	With tie plate
9	PAWL, bottom rear #	BD 0795	Steel	1	MGD 78	With tie plate
10	PAWL, top, front #	BD 0797	Steel	1	MGD 80	
11	PAWL, top, rear #	BD 0799	Steel	1	MGD 81	
12	PIN, axis, bottom pawls #	BD 0805	Steel	1	MGD 83	
13	PIN, axis, top pawls #	BD 0808	Steel	1	MGD 88	
14	PIN, split, levers, feed block #	BD 0835	Steel	1	MGD 108	Spring pin
15	PLATE, connecting, bottom pawls	BD 7967	Steel	1	MGD 120	
16	RIVET	CA 0441	Steel	4	MGD 149	UK Vocab. No., no Aust. listing
17	SLIDE, Mk I* #	BD 0930	Steel	1	MGD 180	
18	SPRING, bottom pawls #	BD 0940	Nickel	1	MGD 185	
19	SPRING, top pawls, No. 1 #	BD 0959	Nickel	1	MGD 198	
20	STRIP	CA 0447	Steel	1	MGD 205	UK Vocab. No., no Aust. listing
21	RIVET	CA 0445	Steel	3	MGD 156	UK Vocab. No., no Aust. listing

\# Component part provided for normal maintenance; held in ordnance store for issue.
‡ Parts marked thus are used elsewhere on the gun; the 'No. off' applies only to each particular Ref. No.

one hand holding down the trigger bar while the other allows the crank handle to move forwards. Once it is in battery, a click should be audible as the sear releases; if it is early (or late), then the lock must be examined very carefully, with the usual remedy being replacement of the sear. Finally, the feedblock should be checked and the pawls examined and lubricated. These move extremely fast as the belt is cycled and if the lower spring is weakened it will not permit the pawl arms to return to the correct position, enabling them to grip the belt and feed the next round. *In extremis* it could enable the belt to simply drop out of the feedblock: embarrassing on the range, but potentially fatal in combat.

After extended firing, checking its tension with a spring balance was recommended; if it fell outside the recommended pull-weight, it must be discarded and replaced. Generally feedblocks were supplied complete, and problems with one simply required a replacement. Faulty ones could be returned to the unit armourer for rectification, as they were difficult to maintain in the field. Having now ascertained that all of the internal parts were correctly set up and lubricated, attention could be turned to the tripod.

The Mk IV tripods themselves were solid items that did not suffer from undue problems in service, other than normal wear. The lower body of the majority of tripods was gunmetal and its elevating gear and shaft were of machined steel, the legs being steel tube, all of which require considerable effort to damage. The three locking handles for the legs were kept oiled but beyond that there was little maintenance required. Slightly more vulnerable were the crosshead and its locating shaft and the directional dial ring which were also made of gunmetal. Common tripod problems normally related to the crosshead, for if the gun was subjected to a heavy blow (shellfire could move a gun several dozen yards), or if it fell heavily, it could bend the crosshead jaws that retain the locking pins, making them impossible to remove. Knocking them out with a hammer was the only solution but the crosshead could be field-repaired, as the manual stated: 'The jaws can be set inwards by means of a rawhide mallet or carefully nipping the crosshead in a vice.' For the gunners, this usually meant hitting it hard with a hammer.

The L-shaped steel pins, which lock the gun body in position, were also vulnerable to being bent or having their surface damaged,

BELOW Instructors devoted much time to explaining the complexities of the firing cycle. This coloured diagram was used for lectures. *(Ian Durrant)*

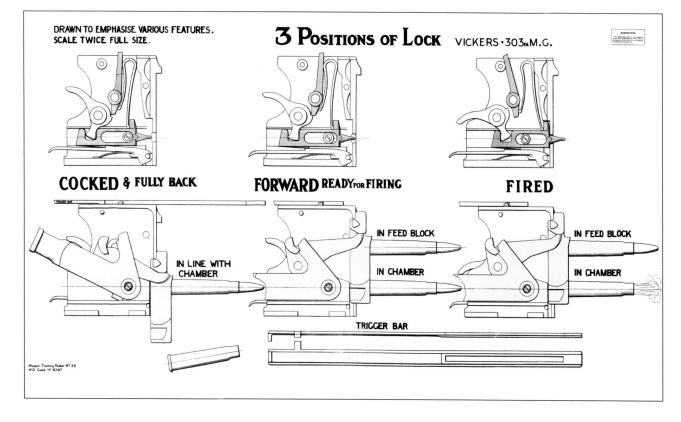

Plate D₁

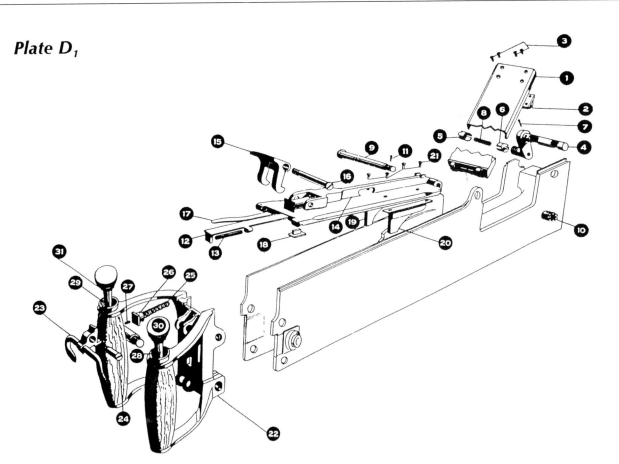

Ref. No.	Designation		Vocab. Number	Mat.	No. Off	Drawing Number	Remarks
	Plate D₁ GUN, MACHINE, VICKERS, .303-in. MK 1 (continued)		BD 7930 GA	...		MGA 790	
D 1	COVER, FRONT, MK I*	#	BD 0651	Steel	1	MGD 35	
2	BRACKET		BD 7673	Steel	1	MGD 12	
3	RIVET		CA 0462	Steel	1	MGD 154A	Also rivet, stop, for Tripod Mk 4B
4	CATCH, Mk I	#	BD 0637	Steel	1	MGD 23	
5	PLUG	#	BD 0872	Steel	1	MGD 126	
6	PLUNGER	#	BD 0875	Steel	1	MGD 128	
7	PIN, keep, split, $^{1}/_{16}$-in. dia. × ½-in. long	#	G1/GA 0799	Steel	1 ‡	...	Malleable, hardware item
8	SPRING, Mk I	#	BD 0943	Spring Steel	1	MGD 187	
9	PIN, joint, cover, Mk IV §		BD 0821	Steel	1	MGD 553	
10	NUT, check, Mk II		BD 0774	Steel	1	MGD 72A	
11	PIN, keep, split, $^{1}/_{16}$-in. dia. × 1-in. long	#	G1/GA 0801	Steel	1 ‡	...	Malleable, hardware item
...	COVER, REAR, MK I		BD 7932	...		MGA 227	
12	BAR, trigger	#	BD 0610	Steel	1	MGD 4	
13	SPRING	#	BD 0960	Spring Steel	1	MGD 199	Hooks over lock trigger
14	COVER		BD 7740	Steel	1	MGD 37	
15	LOCK	#	BD 0765	Steel	1	MGD 70	
16	PIN	#	BD 0810	Steel	1	MGD 90	
17	SPRING	#	BD 0952	Spring Steel	1	MGD 191	
18	STUD	#	BD 0973	Steel	1	MGD 209	
19	RAMP, left		BD 7995	Steel	1	MGD 130	Rivetted in position
20	RAMP, right		BD 7996	Steel	1	MGD 131	Rivetted in position
21	RIVET, ramp		CA 0462	Steel	4	MGD 154A	Also rivet, stop, for Tripod Mk 4B
...	CROSSPIECE, REAR, MK I		BD 0882A	...		MGA 230	
22	BODY		BD 7650	Steel	1	MGD 132	
23	CATCH, safety, Mk I	#	BD 0638	Steel	1	MGD 24	Released with forefingers
24	PIN	#	BD 0828	Steel	1	MGD 103	
25	SPRING, Mk I	#	BD 0953	Spring Steel	1	MGD 192	
26	PISTON		CA 0512	Steel	1	MGD 117	UK Vocab. No., no Aust. listing
27	RIVET		CA 0455	Steel	1	MGD 151	UK Vocab. No., no Aust. listing
28	BOTTLE, oil	#	BD 0622	Steel	2	MGD 8	BD 7657 (UK). Left & right
29	BRUSH, oil, Mk II		BD 4197	Plastic Mld.	2	MGD 463	One brush in each grip
30	HEAD, milled, M.G., Mk I		CA 0532	Steel	2	MGD 53	UK Vocab. No., no Aust. listing
31	WASHER, leather, M.G., Mk I	#	BD 2696	Leather	2	MGD 220	

§ Pin, joint, cover, Mk V is fitted when used with Bracket, dial sight Vocab. BD 0823

preventing insertion. The holes in the brass crosspiece through which they needed to pass eventually became enlarged (these were factory drilled to ⁹⁄₁₆in for the front pin and ⁷⁄₁₆in for the rear), so oversize pins were available. If this did not solve the problem, the holes had to be reamed out and rebushed, a skilled job not possible in the field. More frequently they simply became clogged with dirt or mud, or burred over, making it very difficult to push the pins home. Dirt was cleared by running a patched cleaning rod through (usually one for the service .455in revolver, as all gunners carried one) but burring required careful filing. The threads on the elevating gear were open to the elements and mud and sand were a continual problem. In desert conditions they were kept unlubricated but in western Europe, they were normally greased to repel rainwater, and keep out much of the debris that might accumulate. Nevertheless, they required regular checking and cleaning.

Assuming that inspection found no play on the tripod or crosshead, there still remained the very necessary checking of ancillary items such as water hoses, water tins and, of course, ammunition. The condenser hoses were examined for splits. If the screw-threaded hose boss under the muzzle was damaged, the gun had to be withdrawn from service, as it was soldered to the jacket of early production guns and could not be field repaired. On post-First World War models, however, the boss was retained by two screws and could be replaced. The water cans were initially commercial petrol cans, which were plentiful – just as well, as they were made of thin, soldered tin and inherently fragile. A stack would be accumulated and filled, as shortage of water was a constant concern when in combat. Having a gun cease firing due to lack of coolant was regarded as inexcusable and there are accounts of men using the urine tins from the infantry to top up the water. By all accounts it worked, but smelled highly unpleasant.

With the water supply secure, it was time to look at the last remaining items, the belts and ammunition. As already outlined, belts were a constant cause for concern, as they had to be perfectly loaded, dry and contain ammunition that had not degraded with storage. Machine gunners spent an inordinate amount of time

The hose connector is unscrewed, allowing the water to drain from the jacket. *(Author)*

The brass protector in place to prevent damage to the delicate brass threads. *(Author)*

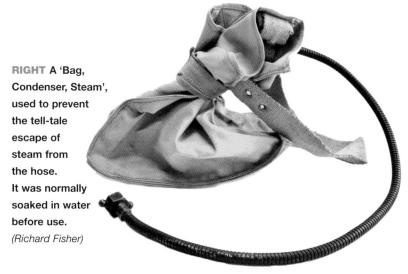

RIGHT A 'Bag, Condenser, Steam', used to prevent the tell-tale escape of steam from the hose. It was normally soaked in water before use. *(Richard Fisher)*

Plate D₂

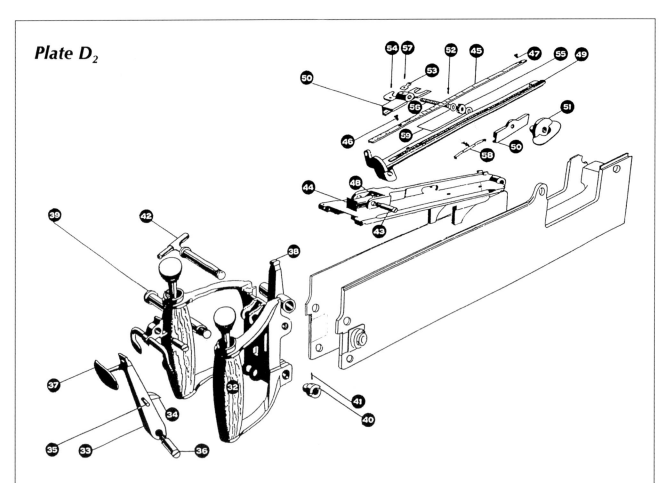

Ref. No.	Designation		Vocab. Number	Mat.	No. Off	Drawing Number	Remarks
Plate D₂	GUN, MACHINE, VICKERS, .303-in. MK 1 *(continued)*		BD 7930 GA			MGA 790	
...	CROSSPIECE, REAR, MK I *(continued)*						
32	GRIP	#	BD 0720	Wood	2	MGD 50	Set of, left or right side fitting
33	LEVER, firing	#	BD 9751	Steel	1	MGD 65	
34	PAWL		...	Steel	1	MGD 79	
35	PIN, axis		...	Steel	1	MGD 85	
36	PIN, axis	#	BD 0806	Steel	1	MGD 84	
37	THUMB-PRESS		...	Steel	1	MGD 210	Firing trigger, with thumbs
38	LEVER, trigger, bar	#	BD 0755	Steel	1	MGD 68	
39	PIN, joint, Mk I	#	BD 0822	Steel	1	MGD 461	
40	NUT, check		...	Steel	1	MGD 73	
41	PIN, keep, split, 1/8-in. dia. x 3/4-in. long	#	G1/GA 0841	Steel	1 ‡	...	Malleable, hardware item
42	PIN, T, fixing, Mk I ¶	#	BD 0840	Steel	1	MGD 462	Long pin is used w/ sight bracket
...	SIGHT, TANGENT, MK III**	#	BD 0927 SA	...	...	...	Mk 4 is BD 0928
43	PIN, axis	#	BD 0807	Steel	1	MGD 87	
44	PISTON	#	BD 0860	Steel	1	MGD 118	
45	PLATE, graduated, No. 2, Mk I	#	BD 0863	Nickel	1	MGD 140	Mk I* is BD 0862, interchangeable
46	SCREW, lower	#	BD 0906	Steel	1	MGD 170	
47	SCREW, upper	#	BD 0907	Steel	1	MGD 171	
48	SPRING, Mk I	#	BD 0958	Spring Steel	1	MGD 197	
49	STEM, Mk II	#	BD 0970	Steel	1	MGD 201	Incorporates offset battle sight
50	SLIDE, tangent sight, Mk II**	#	BD 0935	Steel	1	SAID 2075B	Mk 3 is BD 0936
51	NUT, No. 2	#	BD 0777	Steel	1	SAID 2075B	
52	PIN, keep, split, 1/16-in. dia. x 1/2-in. long	#	G1/GA 0799	Steel	1	SAID 2075B	Malleable, hardware item
53	PILLAR, aperture		...	Steel	1	SAID 2075B	Vocab. No. CA 0522 (UK)
54	RIVET		...	Steel	1	SAID 2075B	Vocab. No. CA 0444 (UK)
55	PINION, Mk II	#	BD 0802	Steel	1	SAID 2075B	
56	SCREW		BD 0910	Steel	1	SAID 2075B	
57	PIN, fixing	#	BD 0818	Steel	1	SAID 2075B	
58	SPRING, friction		BD 0942	Spring Steel	1	SAID 2075B	
59	WASHER		BD 0995	Steel	2	SAID 2075B	

¶ Alternative Mk II measurements for Dial Sights to be shown
\# Component part provided for normal maintenance; held in ordnance store for issue.
‡ Parts marked thus are used elsewhere on the gun; the 'No. off' applies only to each particular Ref. No.

RIGHT **The split-pin is removed to allow the muzzle attachment to be unlocked.** *(Author)*

CENTRE **The attachment is removed with a 120° twist.** *(Author)*

checking ammunition when filling belts, or handling preloaded belts, but by the Second World War the practice of using factory-sealed tins meant that examining preloaded belts was impossible until the moment they were opened. If belts were to be reloaded, the gunners would sit with open boxes of .303in SAA checking rounds individually for signs of corrosion due to water (likely to lead to case separation), improperly inserted primers (which would result in a misfire and stoppage), cartridges that had abnormally thick rims (which would jam in the carrier) or bullets that were damaged or improperly seated.

Aircrew in particular were fixated about the quality of their ammunition, as a stoppage in the air was a serious matter. Pilots often therefore spent hours of spare time obsessively checking their belts for faulty cartridges or links. Once satisfied, the gunners put checked cartridges into a crate or empty ammunition boxes to then be loaded into belts.

The belts themselves were also regularly inspected, as they led a hard life, being dumped on the ground once empty, accumulating dirt and mud as well as becoming damaged by being trodden on. Those with frayed edges, lost feed tabs, bent or missing brass spacers were discarded but still proved useful – they burned well if doused in a little petrol and were often the only source of fire to provide hot water for a crew's tea, which is possibly why they are so scarce today.

Belt-filling machines were vital, so a great deal of time was spent in carrying SAA boxes to and fro and filling the belts, possibly the

RIGHT **Two MMGS soldiers operate the belt loader; one keeps the hopper filled with cartridges, while the other winds the loading handle and ensures the belt feeds smoothly through the bed of the loading tray.** *(IWM FLM 2051)*

Plate E

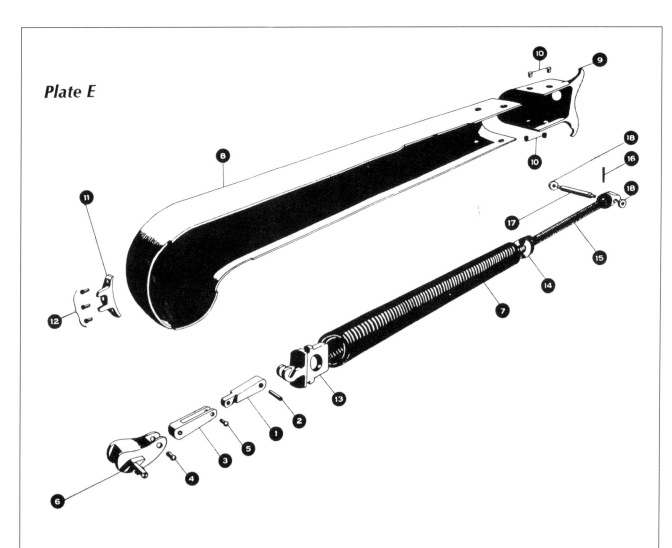

Ref. No.	Designation							Vocab. Number	Mat.	No. Off	Drawing Number	Remarks
Plate E	**GUN, MACHINE, VICKERS, .303-in. MK 1** *(continued)*							BD 7930 GA			MGA 790	
. . .	FUSEE, with CHAIN	.	.	.	.	.	#	BD 0700 A			MGA 231	
. . .	CHAIN							SA			. . .	
E 1	LINK, No. 1	.	.	.	.	.		CA 0482	Steel	1	MGD 27	UK Vocab. No., no Aust. listing
2	PIN	.	.	.	.	.		CA 0484	Steel	1	MGD 102	UK Vocab. No., no Aust. listing
3	LINK, No. 2	.	.	.	.	.		CA 0483	Steel	1	MGD 28	UK Vocab. No., no Aust. listing
4	PIN, fixing	.	.	.	.	.		CA 0484	Steel	1	MGD 96	UK Vocab. No., no Aust. listing
5	RIVET	.	.	.	.	.		CA 0486	Steel	1	MGD 145	UK Vocab. No., no Aust. listing
6	FUSEE, MK I	.	.	.	.	.		BD 7801	Steel	1	MGD 46	
7	SPRING	.	.	.	.	.	#	BD 0944	Spring Steel	1	MGD 188	MGA 382 (UK)
8	BOX, Mk I	.	.	.	.	.	#	BD 0624	Steel	1	MGD 9	Sheet steel
9	LUG, front	.	.	.	.	.		BD 7789	Steel	1	MGD 45	
10	RIVET	.	.	.	.	.		CA 0450	Steel	4	MGD 144	UK Vocab. No., no Aust. listing
11	LUG, rear	.	.	.	.	.		CA 0449	Steel	1	MGD 133	UK Vocab. No., no Aust. listing
12	RIVET	.	.	.	.	.		CA 0451	Steel	3	MGD 150	UK Vocab. No., no Aust. listing
13	HOOK	.	.	.	.	.		BD 7848	Steel	1	MGD 57	
14	NUT	.	.	.	.	.		CA 0524	Steel	1	MGD 427	Mk I* nut (UK)
15	SCREW, adjusting	.	.	.	.	.	#	BD 0901	Steel	1	MGD 169	
16	PIN, stop	.	.	.	.	.		CA 0516	Steel	1	MGD 111	UK Vocab. No., no Aust. listing
17	PIN, vice	.	.	.	.	.		BD 7852	Steel	1	MGD 216	
18	KNOB	.	.	.	.	.		CA 0514	Steel	2	MGD 60	UK Vocab. No., no Aust. listing

\# Component part provided for normal maintenance; held in ordnance store for issue.

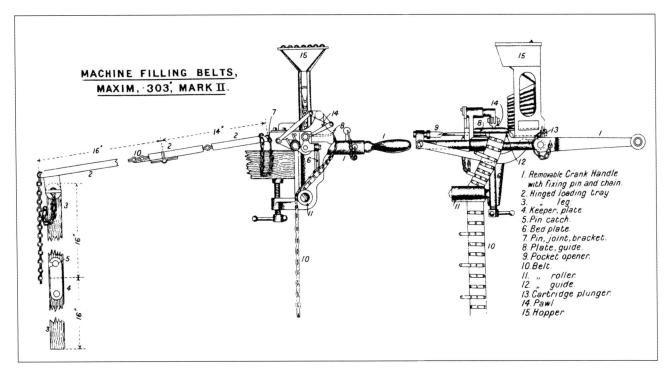

ABOVE A diagram from a Second World War Vickers manual of the belt-loading machine, supplied as part of the Vickers accessories. Although large and heavy it was quite efficient once set up. *(Author)*

nearest Emma Gees got to having to undertake fatigues, as it was heavy, repetitious work. Hand loading belts was possible by pressing the cartridge base on a hard surface to push it into the belt, but it was a very slow procedure and became extremely tiring on the hands and arms. It was best accomplished by relays of gunners if the situation required it. Generally, this was avoided if at all possible and belted ammunition was stockpiled in advance in huge dumps, preferably in a dugout where it was relatively safe from enemy fire and accessible for the gunners, although carrying two boxes of ammunition up a steep flight of stairs several dozen times proved extremely tiring.

It should be pointed out that these checks were generally undertaken once the crew were out of the line and had the time to inspect components at leisure and rectify any problems, often with the assistance of the unit armourer. With the gun now reassembled and operating perfectly, it was time to set it up. The guns and tripods were heavy to carry for any length of time and even assembling one on a range was quite demanding if any distance had to be covered. On a remote firing range often used by the author this meant a quarter-mile walk, inevitably carrying the tripod. Being a large-framed man has its disadvantages. Having to take a gun and its ancillaries over shell-torn ground, sometimes for hundreds of yards required fitness and stamina. Siting the gun was normally determined by the commander in advance so time was not wasted. It had to be out of observation of the enemy, have a clear line of fire to the target area, and there needed to be reasonable access to stockpiled spares, water and ammunition. The crew also required some form of shelter, as they would invariably be targeted by the enemy's artillery, although this was not always possible.

Setting up a gun to fire was a rapid process for a well-drilled crew. The No 1 gun team member placed the tripod in position, ensuring it was firmly seated on the ground, the feet weighted with sandbags if necessary. The No 2 carried the gun and placed it into the crosspiece, No 1 ensuring the locking pins were properly inserted and secured into position. The gun was then mounted on its tripod and No 3 placed an open can of ammunition next to the right side of the gun, along with the water tin and condenser hose, then he moved to its left side, attaching the hose to the gun. He then decamped, ready to bring up more ammunition, water or spare barrels.

No 1 had by now seated himself behind the gun, and No 2 was lying with the ammunition box in front of him. In the interim, the other members of the gun crew would have brought

Plate F

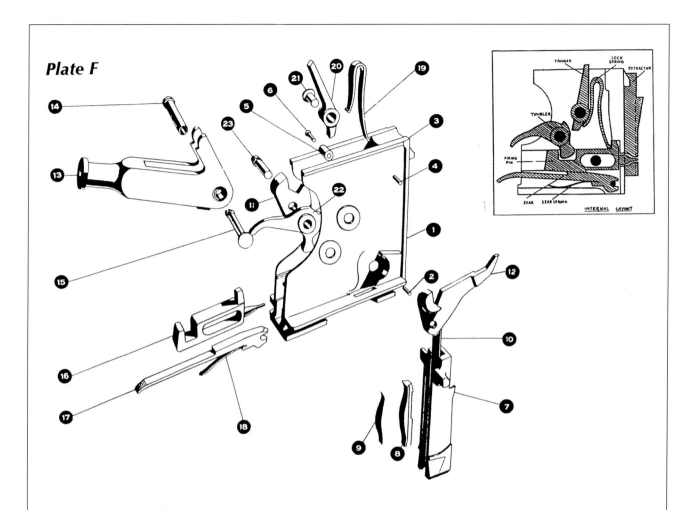

Ref. No.	Designation		Vocab. Number	Mat.	No. Off	Drawing Number	Remarks
Plate F	GUN, MACHINE, VICKERS, .303-in. MK 1 (continued)		BD 7930 GA			MGA 790	
...	LOCK, BREECH, MK I	#	BD 0760 A			MGA 224	
F 1	CASING, lock		BD 7700	Steel	1	MGD 21	
2	PIN, axis, sear		BD 7921	Steel	1	MGD 86	
3	SEATING, spring, lock, No. 2		CA 0492	Steel	1	MGD 176A	UK Vocab. No., no Aust. listing
4	RIVET		CA 0490	Steel	1	MGD 147	UK Vocab. No., no Aust. listing
5	TUBE, distance		CA 0493	Steel	1	MGD 213	UK Vocab. No., no Aust. listing
6	RIVET		CA 0489	Steel	1	MGD 143	UK Vocab. No., no Aust. listing
7	EXTRACTOR, No. 2, Mk I	#	BD 0691	Steel	1	MGD 43	
8	GIB, No. 3	#	BD 0712	Steel	1	MGD 47	
9	SPRING	#	BD 0946	Spring Steel	1	MGD 189	
10	COVER	#	BD 0652	Steel	1	MGD 36	
11	LEVER, left, No. 2	#	BD 0746	Steel	1	MGD 63	
12	LEVER, right, No. 2	#	BD 0748	Steel	1	MGD 64	
13	LEVER, side		...	Steel	1	MGD 66	Fixes over connecting rod
14	BUSH, axis	#	BD 0628	Steel	1	MGD 15	
15	PIN, split	#	BD 0830	Steel	1	MGD 104	Spring pin
16	PIN, firing, No. 2	#	BD 0813	Steel	1	MGD 92	
17	SEAR	#	BD 0920	Steel	1	MGD 177	
18	SPRING	#	BD 0954	Spring Steel	1	MGD 193	
19	SPRING, lock, No. 2	#	BD 0949	Spring Steel	1	MGD 190	
20	TRIGGER	#	BD 0983	Steel	1	MGD 211	Tripped by slide, trigger bar
21	PIN	#	BD 0842	Steel	1	MGD 114	
22	TUMBLER	#	BD 0987	Steel	1	MGD 215	
23	PIN	#	BD 0844	Steel	1	MGD 115	

\# Component part provided for normal maintenance; held in ordnance store for issue.

up additional ammunition, spare barrels and anything else required. Depending on the type of fire, direct or indirect, the gun commander may have already set up an observation post with rangefinder, binoculars, maps and so on. In combat, the lock would normally already have been checked and be in the gun, so No 2's next task was to take the brass tab of the loaded belt, pass it into the feedblock and pull it sharply to the left, keeping tension on it as the No 1 pulled the crank handle back twice.

The first pull would extract a round from the belt, which is then chambered as the lock runs forward into battery, the second pull enabling another cartridge to be gripped in the feedblock ready to extract. Only after both cocking actions would No 2 release the tension on the belt. If the target was visual, the range was called out by the commander and the rear sight adjusted accordingly. If indirect fire was to be employed, then degree co-ordinates and angle of fire would be given to the No 2, who set the barrel according to the 360° graduated ring on the crosspiece, and the No 1 adjusted the tripod elevating gear to the required angle, taken from the clinometer held on top of the receiver or from the dial sight.

There was a correct method of holding the spade-grips to ensure proper control when firing. With both forefingers resting over the top of each spade-grip, the second (longest) fingers hook underneath the safety bar and the thumbs press against the spade-grip. It sounds uncomfortable, but is actually quite instinctive and results in a firm but not unduly tiring grip. Upon the gun commander's shout of 'Fire' the second fingers lift the safety bar, the spade-grip is pressed, and all of Hiram Maxim's engineering genius bursts to life.

Describing firing a Vickers for the first time is difficult. There is a hesitancy when the trigger is pressed, and the thought goes through one's mind that quite possibly nothing is going to happen. But it does, with sudden energy; the gun begins to rock to and fro, albeit in a surprisingly controlled manner, and the belt runs through the feedblock at an incredibly rapid rate. The crank handle flies to and fro, reinforcing the instructor's warning never to place a hand near it. Above all else, the rhythmical sound of the gunfire is

ABOVE The correct firing grip for the Vickers. The thumbs press the trigger, the second fingers push the safety bar upward to unlock it, while the forefingers hook over the grip bars to control traverse. *(Author)*

almost hypnotic, the sound is very much what everyone expects a machine gun to sound like, a series of sharp, staccato 'Rat-a-tat, rat-tat-tat-tat-tats', and quite unlike the slower more measured: 'Tac-tac-tac. Tac-tac, tac-tac-tac', of the Maxim gun, which sounds more like a large hammer tapping on a metal plate. Their distinctive sounds enabled soldiers to determine exactly which side was firing.

Firing the Vickers is quite different to the experience of shooting any early hand-cranked machine gun, such as the Gatling. These are extremely physical to manipulate, as well as having an uneven, hesitant rate of fire. This is due partly to the need for one's arm to rotate the heavy firing handle and also because of misfires, which create unexpected pauses when firing. They feel somewhat like trying to keep a temperamental old static engine working, with its coughs and splutters. But the Vickers has been likened to the equivalent of a modern locomotive engine, with its precise, oiled internals, the lock and conrod all running smoothly forwards and backwards on polished steel surfaces.

What is surprising is the sheer violence of the impact of bullets exiting the muzzle at 500r/pm. The level of fire a single gun can lay down is breathtaking and appalling in equal measure as the target simply vanishes in 5ft high pillars of dirt, sand and earth, as though a ferocious whirlwind was enveloping it. In reality, none of these things were the concern of a machine gunner, who was dispensing death on an industrial scale in the hope of defeating an attacking enemy and staying alive in the

Plate G

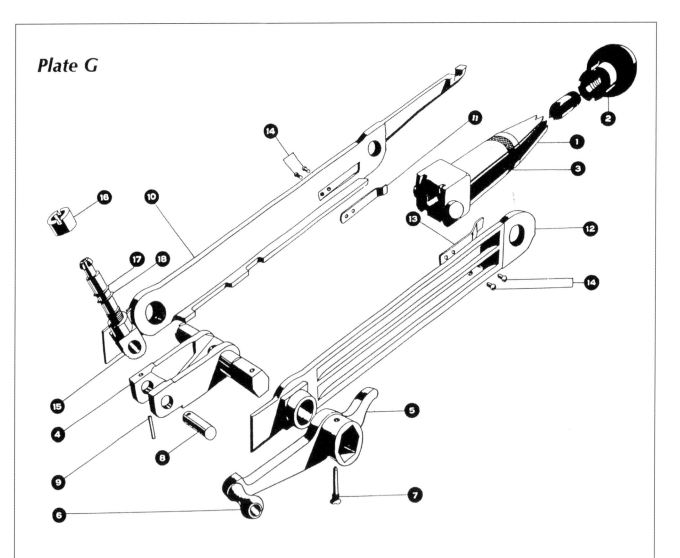

Ref. No.	Designation	Vocab. Number	Mat.	No. Off	Drawing Number	Remarks
Plate G	GUN, MACHINE, VICKERS, .303-in. MK 1 (continued)	BD 7930 GA			MGA 790	
...	SIDE-PLATES & CRANK	A			MGA 225	
G 1	BARREL, Mk II #	BD 0604	Steel	1	MGD 5	Mk I is lighter, no muzzle thread. Barrel life: 10,000-12,000 rounds
2	CUP, muzzle, attachment, ball, Mk II #	BD 0660	Steel	1	MGD 39	Mk I cup is slide fit
3	PACKING, asbestos † #	BD 0790	Asbestos	1	...	String
4	CRANK, Mk IA ¤	BG 0230	Steel	1	MGD 38	
5	HANDLE #	BD 0730	Steel	1	MGD 52	
6	KNOB	CA 0500	Steel	1	MGD 59	UK Vocab. No., no Aust. listing
7	PIN, fixing #	BD 0816	Steel	1	MGD 94	
8	PIN #	BD 0811	Steel	1	MGD 91	
9	PIN, fixing #	BD 0817	Steel	1	MGD 95	
10	PLATE, side, left, No. 1 Mk I #	BD 0864	Steel	1	MGD 124	
11	SPRING #	BD 0956	Spring Steel	1	MGD 194	
12	PLATE, side, right, No. 1 Mk I #	BD 0865	Steel	1	MGD 125	
13	SPRING #	BD 0957	Spring Steel	1	MGD 195	
14	RIVET, spring, side plates #	BD 0890	Steel	4	MGD 152	
15	ROD, connecting, Mk I #	BD 0897	Steel	1	MGD 165	Connects onto breech lock
16	NUT, adjusting #	BD 0770	Steel	1	MGD 71	
17	WASHER, adjusting, No. 1, .003-in. #	BD 0993	Steel	3	MGD 217	Headspace
18	WASHER, adjusting, No. 2, .005-in. #	BD 0994	Steel	3	MGD 218	Headspace

† As required.
¤ Or Mk I, BD 0657.
\# Component part provided for normal maintenance; held in ordnance store for issue.

WORKING POSITIONS OF LOCK.

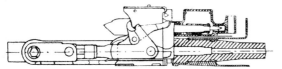

Lock fully home and just fired. Extractor engaging with empty case in chamber and cartridge in feed block.

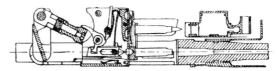

Lock and barrel recoiling. Extractor withdrawing empty case from chamber and a cartridge from the feed block, firing pin cocked and safety sear engaging.

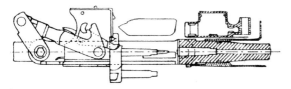

Lock in nearly fully recoiled position. Barrel returning. Extractor down, brings cartridge in line with chamber and empty case either falls off or is pushed off when extractor rises.

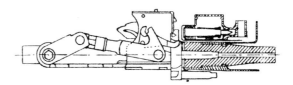

Lock returning, barrel home, extractor being raised by levers, leaving empty case to be ejected, cartridge in chamber, and about to engage with another in the feed block.

[See pp. 69 and 71 (Secs. 29 and 30).]

PLATE V.

ABOVE The firing cycle of the lock and conrod, which are easier to comprehend with a demonstration gun and lock at hand. *(Ian Durrant)*

process. Unsurprisingly, surrender by machine gunners was rarely acceptable to the enemy.

Topping up the water was a continual affair, requiring 1½ pints after 500 rounds of rapid fire and this could be done with the gun in action, although care needed to be taken to prevent scalding from the steam. In fact, despite the careful packing of the barrel glands, in cool or cold weather steam escaped in large amounts from the muzzle and also from the water tin, pinpointing the gun's location to an observant enemy. Inevitably, there would be a requirement to change a barrel, and for this the gun had to cease fire. The expertise of a crew determined how quickly this could be done. Bear in mind every single part of the gun would by now be extremely hot and capable of inflicting second-degree burns on bare skin.

Removing the rear locking pin on the cradle enabled the gun to be raised at the breech and tilted downwards. This meant that the water in the jacket did not drain from the rear as soon as the barrel was withdrawn. The No 1 then removed the lock, the fusee spring cover and the fusee spring. He then unscrewed the T-shaped rear locking pin from the left side of the receiver, which enabled the grips to be swung downward, still hinged on their lower retaining pin.

Meanwhile, No 2, having unlocked the muzzle booster, was feverishly using a mallet

BELOW Gunners use an old saucepan to top up the water jacket on a Vickers in the front lines. Despite the official issue of equipment, they frequently had to employ whatever was at hand. *(Author)*

Plate H

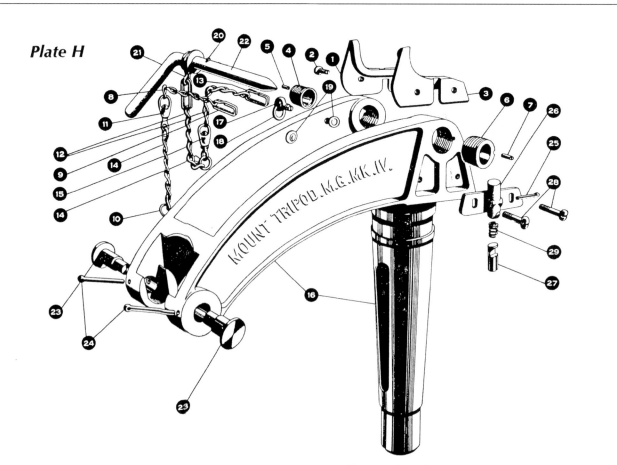

Ref. No.	Designation		Vocab. Number	Mat.	No. Off	Drawing Number	Remarks
Plate H	MOUNTING, TRIPOD, .303-in. M.G., MK IV B		BD 2226 GA	...		DD(E) 2138	UK Drawing as DD(R) 2138
...	CROSSHEAD, 'B'	#	BD 2243 MA	...	1	MGA 1863	
H 1	BLOCK, centering gun, left	#	BD 2227	Bronze	1	MGD 1883	
2	SCREW, securing	#	BD 8890	Steel	1	MGD 1937	
3	BLOCK, centering gun, right	#	BD 2228	Bronze	1	MGD 1884	
4	BUSH, left	#	BD 8896	Steel	1 ‡	MGD 1888	
5	PIN, securing		BH 2344	Steel	1 ‡	MGD 1921	Updated to BH 2309
6	BUSH, right	#	BD 8897	Steel	1 ‡	MGD 1888	
7	PIN, securing		BH 2344	Steel	1 ‡	MGD 1921	Updated to BH 2309
...	CHAIN, ELEVATING SCREW, MK II	#	BD 2236 SA	...	1	MGA 1859	
8	CHAIN (6 links)		BH 2305	Steel	2 ‡	MGD 1891	
9	LINK, split	#	BD 2268	Spring Steel	1 ‡	MGD 1908	
10	RING		BH 2306	Steel	2 ‡	MGD 1932	
11	SWIVEL		BH 2307	Steel	1 ‡	MGD 1957	
...	CHAIN, PIN, JOINT, MK II	#	BD 2238 SA	...	1	MGA 1861	
12	CHAIN (6 links)		BH 2305	Steel	2 ‡	MGD 1891	Same as Ref. No. 8
13	LINK, split		BD 2268	Spring Steel	2 ‡	MGD 1908	
14	RING		BH 2306	Steel	1 ‡	MGD 1932	Same as Ref. No. 10
15	SWIVEL		BH 2307	Steel	1 ‡	MGD 1957	Same as Ref. No. 11
16	CROSSHEAD		BH 2304	Iron	1	MGA 1892	Malleable cast iron
...	EYE & RING		BH 2310 SA	...	2	MGA 1866	
17	EYE		BD 8903	Steel	2	MGD 1894	
18	RING		BH 2306	Steel	2 ‡	MGD 1932	Same as Ref. No. 10
19	WASHER	#	BD 2337	Steel	2	MGD 1962	
...	PIN, JOINT, CROSSHEAD, MK II B		BD 2281	...	1	MGA 1874	
20	FEATHER		...	Steel	1	MGD 1895	
21	LOOP	#	BD 2269	Steel	1	MGD 1906	
22	PIN		...	Steel	1	MGD 1918	
23	PIN, tumbler	#	BD 2290	Steel	1	MGD 1924	
24	PIN, keep, split, ⅛-in. x 2½-in.	#	G1/GA 0845	Steel	2	...	Malleable, hardware item
...	POINTER, DIAL DIRECTION, MK II		BD 2301 SA	...	1	MGA 1876	
25	PIN, keep, split, ⅛-in. x 1-in.	#	G1/GA 0842	Steel	1	...	Malleable, hardware item
26	PLATE	#	BD 2295	Bronze	1	MGD 1929	
27	PLUNGER, indicator	#	BD 2298	Bronze	1	MGD 1930	
28	SCREW	#	BD 2318	Steel	2	MGD 1943	
29	SPRING, plunger, indicator	#	BD 2331	Spring Steel	1	MGD 1954	

and 'C' spanner to undo the muzzle cup. How long this took depended entirely on how reluctant it was to unscrew. Once off, the gunner held the crank handle firmly in his right hand and pulled it rearwards, sliding out both sideplates along with the crank, its conrod and the barrel as a single assembly.

A cork, screwdriver or spent case wrapped in flannelette was quickly used to plug the hole left by the barrel in the end cap of the water jacket and the hose was removed from the tin and wrapped around it as the coiled hose would not allow water to escape.

The sideplates were sprung apart and the trunnions of the barrel would drop free of their locating holes; the barrel was then discarded and a new one inserted. This had to be pre-wrapped with oiled asbestos packing at the breech end, which would have been done prior to going into action. As long as it wasn't obviously leaking, the front gland would remain in situ. Depending on its condition, a new muzzle cup may be fitted. The entire process was then reversed and assuming the cup had

LEFT The 'T' pin at the left rear of the receiver is unscrewed and the grips dropped on their lower hinge, exposing the trigger bar and spring. *(Author)*

not been unusually reluctant to unscrew, a skilled crew could manage this complex drill in around two minutes. The author has replicated this, but only on a perfectly clean and oiled gun, that had not been fired. His first attempt with a

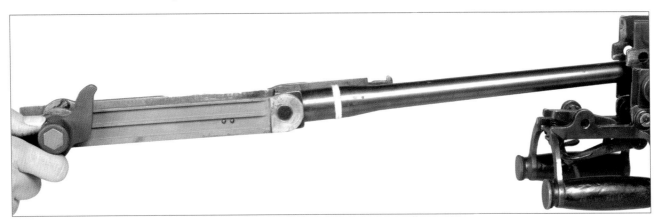

ABOVE Holding the crank handle, the entire mechanism, sideplates and barrel can be withdrawn from the receiver as one unit. Note the asbestos packing on the rear of the barrel. *(Author)*

RIGHT The sideplates are separated and the barrel drops free. Its securing trunnions are visible, as are the sideplate springs which act as guides for the carrier/extractor. The entire gun can now be cleaned, and a new barrel attached on reassembly. *(Author)*

Plate J

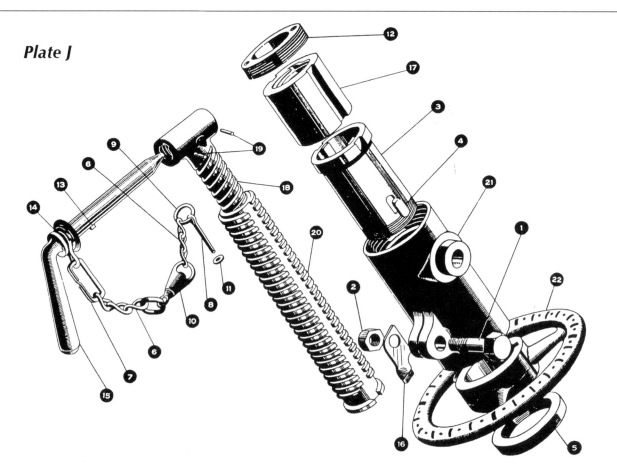

Ref. No.	Designation			Vocab. Number	Mat.	No. Off	Drawing Number	Remarks
Plate J	MOUNTING, TRIPOD, .303-in. M.G., MK IV B *(continued)*			BD 2226 GA			DD(E) 2138	
...	CROSSHEAD, 'B' *(continued)*							
...		GEAR, ELEVATING 'B'		BD 2256 A		1	MGA 1865	
J 1			BOLT, jamming #	BD 2230	Steel	1	MGD 1887	UK list changed from BD 2480
2			NUT	...	Steel	1	MGD 1909	
3			BUSH, wheel, elevating #	BD 2231	Bronze	1	MGD 1889	
4			FEATHER #	BD 8898	Steel	1	MGD 1898	
5			NUT #	BD 2275	Brass	1	MGD 1910	
...			CHAIN, MK II #	BD 2240	...	1	MGA 1860	
6			CHAIN (3 links)	...	Steel	2	MGD 1890	
7			LINK, split #	BD 2268	Spring Steel	1 ‡	MGD 1908	
8			PIN, keep, split, ⅛-in. x 1½-in. #	G1/GA 0843	Steel	1 ‡	...	Malleable, hardware item
9			RING	...	Steel	2 ‡	MGD 1932	
10			SWIVEL	...	Steel	1 ‡	MGD 1957	
11			WASHER #	BD 2338	Steel	1	MGD 1961	
12			NUT, tumbler	...	Brass	1	MGD 1915	
...			PIN, JOINT, ELEVATING GEAR, MK II B #	BD 2284 SA	...	1	MGA 1875	
13			FEATHER	...	Steel	1	MGD 1896	
14			LOOP #	BD 2270	Steel	1	MGD 1907	
15			PIN	...	Steel	1	MGD 1919	
16			POINTER #	BD 2302	Steel	1	MGD 1931	
...			SCREW, ELEVATING #	BD 2316 SA	...	1	...	Vertical adjustment
17			NUT	...	Steel	1	MGD 1911	
18			SCREW, inner	...	Steel	1	MGD 1939	RH thread outside
19			PIN, stop	...	Steel	2	MGD 1923	
20			SCREW, outer	...	Steel	1	MGD 1940	Interrupted LH thread
21			TUMBLER #	BD 8966	Bronze	1	MGD 1960	UK list is BD 8925
22			WHEEL #	BD 2351	Iron	1	MGD 1967	Malleable cast iron. Segment slots were cut into upper surface, each segment representing ten minutes of elevation; dimples in between being five minutes. For night time operation.

\# Component part provided for normal maintenance; held in ordnance store for issue.
‡ Parts marked thus are used elsewhere on the gun; the 'No. off' applies only to each particular Ref. No.

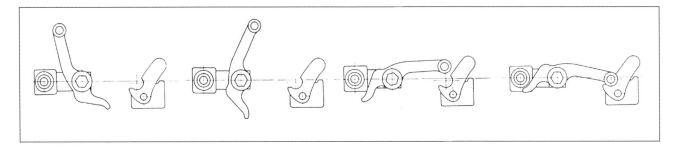

working gun, with two others helping, took ten minutes and the loss of all the coolant water.

All crews were trained to deal with the inevitable stoppages and the Vickers manual illustrates the four most basic stoppages, which crews were trained to identify at a single glance by looking at the position of the crank handle. There were several possible reasons for each type of stoppage, which determined why the handle had stopped in the position it did, but the only way to find out the exact cause was to open the two top covers and examine the mechanism. The most common faults (there were others) were as follows:

First position of crank handle
- Weak charge
- Weak or broken gib spring
- Too heavy fusee spring
- Grit in working parts or insufficient oil
- Excessive gland packing
- Worn barrel
- Cartridges too tight in belt
- Friction due to freezing.

Second position
- Damaged cartridge
- Separated case.

ABOVE The crank handle shown in the one to four positions that denote stoppages. Every Vickers gunner had to learn these by heart, and in training were blindfolded to ensure they could clear a stoppage by touch. It was particularly important for pilots. *(Author)*

LEFT The elevation gear on the Mk IV tripod. The wheel raises or lowers the gun. The locking pin has been removed from the top mount to release the receiver from the crosshead. *(Author)*

FAR LEFT The two pintle securing pins both have machined lugs that ensure once in position they cannot slip out unless the lever is turned through 180°. *(Author)*

LEFT The grips each contain an oil reservoir and application brush. *(Author)*

Plate K

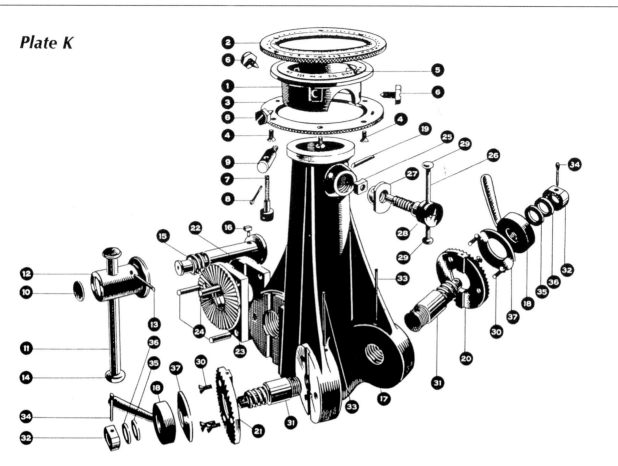

Ref. No.	Designation		Vocab. Number	Mat.	No. Off	Drawing Number	Remarks
	Plate K MOUNTING, TRIPOD, .303-in. M.G., MK IV B *(continued)*		BD 2226 GA			DD(E) 2138	
...	DIAL, DIRECTION, MK II	#	BD 2251 A	...	1	MGA 1864	
K 1	BODY			Iron	1	MGD 1886	Malleable cast iron
...	RING, ADJUSTABLE		BD 2310 SA	...	1		
2 & 3	RINGS, graduated & plain			Bronze	1 ea.	MGD 1933 & 1934	Two rings
4	SCREW, fixing	#	BD 2317	Brass	6	MGD 1942	
5	SCREW, zero	#	BD 2322	Steel	1	MGD 1946	
6	SCREW, set	#	BD 2320	Steel	3	MGD 1944	
7	SCREW, set, rings	#	BD 2321	Steel	1	MGD 1945	
8	PIN, keep, split, 1/16-in. x ½-in.	#	G1/GA 0799	Steel	1		Malleable, hardware item
9	STUD, set screw, rings	#	BD 2336	Steel	1	MGD 1956	
...	NUT, JAMMING, REAR LEG		BD 2276 A	...	1	MGA 1873	
10	DISC		BH 2326	Steel	1	MGD 1893	
11	HANDLE	#	BD 2261	Steel	1	MGD 1900	Sliding bar
12	NUT		CA 0834	Steel	1	MGD 1912	UK Vocab. No., no Aust. listing
13	PIN, fixing	#	BD 9112	Steel	1 ‡	MGD 1916	
14	WASHER, handle	#	BD 9013	Steel	1	MGD 1963	
15	PIN, joint, rear leg	#	BD 2286	Steel	1	MGD 1920	
16	FEATHER	#	BD 9107	Steel	1	MGD 1897	
...	SOCKET		A		1	MGA 1879	
17	BODY			Iron	1	MGD 1952	Malleable cast iron
18	HANDLE, jamming, front legs	#	BD 2260	Steel	2	MGD 1899	Fixed handle, not sliding
19	PIN, nut, screw, jamming crosshead §	#	BD 9112	Steel	1 ‡	MGD 1916	
20 & 21	PLATES, clutch, front, left & right	#	BD 8927	Steel	1 ea.	MGD 1925 & 1926	Two plates
22 & 23	PLATES, clutch, rear, left & right	#	BD 8928-9	Steel	1 ea.	MGD 1927 & 1928	Two plates
24	RIVET, plate, clutch, rear			Steel	3	MGD 1935	
...	SCREW, CLAMP, CHECKING TRAVERSE	#	BD 2315 SA	...	1	MGA 1862	
25	BLOCK	#	BD 2229	Brass	1	MGD 1885	
26	HANDLE	#	BD 8916	Steel	1	MGD 1901	Sliding bar
27	NUT	#	BD 2277	Brass	1	MGD 1913	
28	SCREW			Steel	1	MGD 1938	
29	WASHER, handle	#	BD 9021	Steel	2	MGD 1964	
30	SCREW, fixing, plates, clutch, front	#	BD 8943	Steel	6	MGD 1941	
31	STUD, joint, front legs	#	BD 2335	Steel	2	MGD 1955	
32	NUT	#	BD 8951	Steel	2	MGD 1914	
33	PIN, fixing			Steel	2	MGD 1917	
34	PIN, keep, split, ⅛-in. x 1½-in.	#	G1/GA 0843	Steel	2 ‡	...	Malleable, hardware item
35 or 36	WASHER, stud, joint, No. 1	#	BD 2340	Steel	2	MGD 1966	Alternative washer No. 2, 1/16-in.
37	SPRING, disc	#	BD 2330	Spring Steel	2	MGD 1953	

§ To be demanded as Pin, fixing, nut, jamming, rear leg, BD 9112

Third position
- Cross-fed cartridge
- Lock friction
- Bent brass strip
- Torn or worn belt
- Loose belt pockets
- Belt not in line with feedblock
- Thick-rimmed cartridge.

Fourth position
- Defective ammunition
- Broken or damaged firing pin
- Broken lock spring
- Empty belt pocket.

There were also special circumstances that could cause a stoppage, and gunners were expected to be able to diagnose these quickly: most revolved around the lock and extractor, or more rarely a broken fusee spring or damaged muzzle cup. Rectification was an immediate replacement of the part. Of the above, the most protracted to deal with was a separated case, where the extractor had torn off the base of a chambered cartridge, then fed a live round on top of it. In this case, both top covers must be opened, the lock and feedblock removed and the live round removed with a screwdriver, pliers or cleaning rod. Then the 'Plug, Clearing Gun, Vickers .303' would be inserted into the chamber. This was a solid steel, cartridge-shaped rod mounted on a hinged handle and when driven into the jammed cartridge case with several hard blows, should grip it firmly enough to enable it to be levered out – usually at least. In some instances it failed to do so, and the only solution then was to replace the barrel. In fairness, it was a rare occurrence, and most stoppages could be quickly dealt with.

One final question is worth looking at, and that was the accuracy of a Vickers gun. Naturally, the subject of accurate shooting for a medium machine gun must be put into perspective. It was not an infantry rifle and was not expected to perform as such; in effect a machine gun acts much like a giant shotgun, its bullets dispersing more as the range increases. But it must be able to strike an aimed target at normal combat distances, and as long as the tripod was sufficiently firmly planted, the Vickers could be surprisingly precise. Testing accuracy in a conventional way by counting bullet strikes on a target bullseye was pointless with a machine gun at close range, as after 20 or so shots there would simply be no target left to examine. It is possible on a 6ft square target at 300yd to shoot the centre out of it with a single belt of cartridges, albeit with some bullet dispersion due to crosswinds or tripod movement. At 500yd the target could be cut in two by using a very tiny amount of traverse. In combat, assuming correct range estimation, a platoon of advancing soldiers could be repeatedly hit at up to 600yd by a gun set in enfilade with virtually no traversing. The range officer at one military range used by the author at 500yd with two Vickers guns firing eventually asked for a ceasefire as there was simply no trace of targets left, only two shattered wooden supports, and the banking was becoming seriously eroded by the bullet impact. The utter impossibility of leaving the safety of a trench and advancing more than a few yards into such a maelstrom quickly becomes self-evident, and the bravery of the men who did so is even more astonishing.

ABOVE The lock cycle required two cartridges to be held by the carrier, one to be inserted into the chamber, the second to be gripped while still in the belt and withdrawn as the gun recoiled. This is the reason the crank handle must be operated twice. The lock pictured shows the position 'Forward, ready for firing', as in the previous diagram. *(Author)*

LEFT A detached feedblock with belt and cartridges in situ. The base of the visible cartridge would be gripped by the carrier and is the upper one shown in the previous photograph. *(Author)*

Plate L

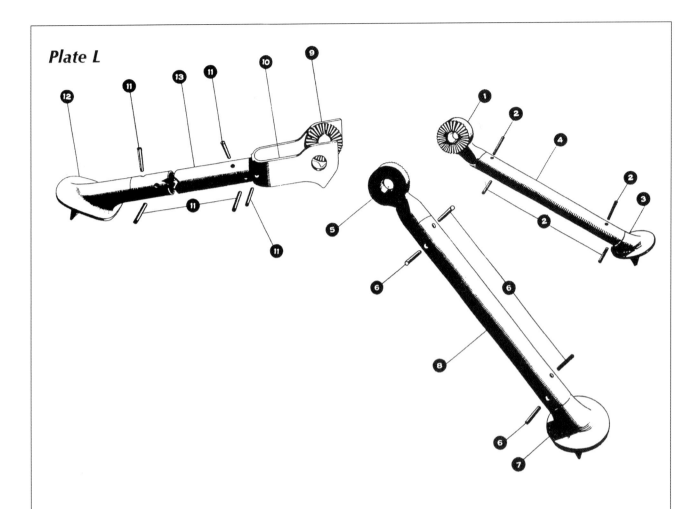

Ref. No.	Designation							Vocab. Number	Mat.	No. Off	Drawing Number	Remarks
Plate L	**MOUNTING, TRIPOD, .303-in. M.G., MK IV B** *(continued)*							BD 2226 GA			DD(E) 2138	
...	LEG, FRONT, LEFT, MK I	.	.	.	.	.	#	BD 2265 A	...	1	MGA 1867	Mk 2 pattern for India only
L 1	JOINT					.		BH 2303	Steel	1	MGD 1902	
2	RIVET					.		BH 2314	Steel	4 ‡	MGD 1936	
3	SHOE					.		BD 8944	Iron	1	MGD 1947	Malleable cast iron, ground spike
4	TUBE					.		BD 9001	Steel	1 ‡	MGD 1958	
...	LEG, FRONT, RIGHT, MK I					.	#	BD 2266 A	...	1	MGA 1868	
5	JOINT					.		BH 2315	Steel	1	MGD 1903	
6	RIVET					.		BH 2314	Steel	4 ‡	MGD 1936	Same as Ref. No. 2
7	SHOE					.		BD 9002	Iron	1	MGD 1948	Malleable cast iron, ground spike
8	TUBE					.		BD 9001	Steel	1 ‡	MGD 1958	Same as Ref. No. 4
...	LEG, REAR, MK I					.	#	BD 2267	...	1	MGA 1869	
9	JOINT, left					.		BH 2316	Steel	1	MGD 1904	
10	JOINT, right					.		BH 2317	Steel	1	MGD 1905	
11	RIVET					.		BH 2314	Steel	5 ‡	MGD 1936	Same as Ref. No. 2
12	SHOE					.		BH 2319	Iron	1	MGD 1949	Malleable cast iron, ground spike in each bottom plate
13	TUBE					.		BH 9001	Steel	1	MGD 1958	Same as Ref. No. 4

\# Component part provided for normal maintenance; held in ordnance store for issue.
‡ Parts marked thus are used elsewhere on the gun; the 'No. off' applies only to each particular Ref. No.

Plate N

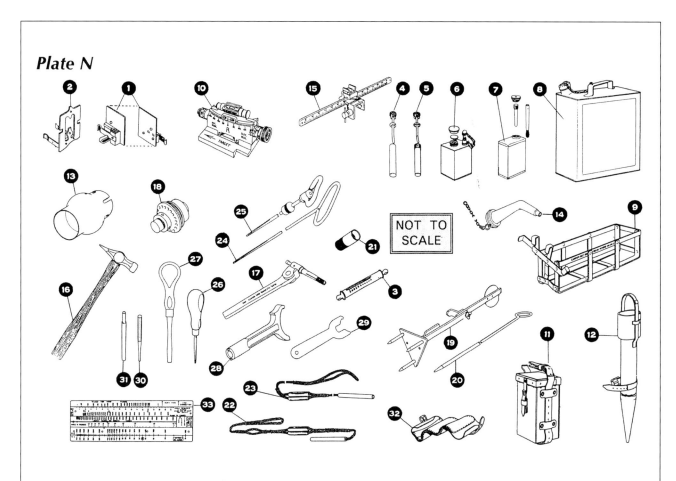

Ref. No.	Designation		Vocab. Number	Mat.	No. Off	Drawing Number	Remarks
Plate N	**ANCILLARY EQUIPMENT, VICKERS M.G., .303-in.**		Sample list only, incomplete				
N 1	Night Sight, back, Vickers, M.G., .303-in., Mk 2 .		BD 2613	Assembly	1	SAID 2212A	Not luminous type
2	Night Sight, fore, Vickers, M.G., .303-in., Mk 1 .		BD 2623	Assembly	1	DD(E) 185A	Not luminous type
3	Balance, spring, M.G., Mk 1	#	BD 0035	Assembly	1	. . .	Brass outer tube
4	Bottle, oil, Mk 4	#	BA 0053	Brass	1	SAID 939A	Also for other .303 arms
5	Bottle, oil, Mk 5	#	SM 520	Plastic	1	DD(E) 2562	Also for other .303 arms
6	Can, half-pint, Mk 2	#	BD 0132	Assembly	1	SAID 1689	For turpentine or oil
7	Can, oil, M.G. Mk 1	#	BD 0133	Assembly	1	SAID 999A	
8	Can, water, 2 gallon	#	LV6/MT1/37556W		1	. . .	General store item, also for fuel
9	Carrier, belt, box, .303-in. Vickers, M.G., Mk 3 .		BD 0136	Steel	1	DD(E) 2428	Clamps onto the gun
10	Clinometer, Vickers, .303-in. M.G., cased, Mk 2 .	#	OS 331GA	Assembly	1	DD(E) 3039	Optical store, for long range firing
11	Case, spare parts, & tools, Vickers, M.G., Mk 1**		BD 0172	Leather	1	DD(E) 927A	With strap & buckles
12	Case, spare barrel & cleaning rod, Vickers, M.G., Mk 2		BD 0165	Leather	1	SAID 1768	With strap & buckles
13	Eliminator, flash & blast deflector, Vickers, M.G., Mk 1	#	BD 6369	Steel	1	DD(E) 2510	
14	Filler, M.G., Mk I		BD 0260	Metal	1	SAID 2052C	For use with condenser can
15	Foresight bar, deflection, Mk 1 .		BD 0270	Steel	1	SAID 1702D	For re-laying the gun
16	Hammer, M.G., Mk 1	#	BD 1000	Assembly	1	SAID 1399A	Wooden handle, flat head
17	Plug, clearing, .303-in., M.G., Mk 1	#	BD 2503	Steel	1	SAID 1256	Ruptured case extractor
18	Muzzle attachment, blank, Mk 1		BD 7897	Steel	1	DD(E) 113	Threaded adjustment
19	Post, aiming, M.G., Mk 3	#	BD 2522	Steel	1	DD(E) 1835	Three-pronged ground spike
20	Post, zero, M.G., Mk 1	#	BD 2516	Steel	1	SAID 1934	
21	Protector, muzzle, Vickers, .303-in., M.G., Mk 2 .	#	BD 9395	Brass	1	DD(E) 2892	Mk I was brass with clamp screw
22	Pullthrough, double, Mk 1A	#	BA 0515	Assembly	1	SAID 287	Also for other .303 arms
23	Pullthrough, single, Mk 4A	#	BA 0517	Assembly	1	SAID 257	Also for other .303 arms
24	Rod, cleaning, .303-in., M.G., Mk 5	#	BD 9225	Steel	1	DD(E) 2686	
25	Rod, cleaning, .303-in., M.G., Mk 2A	#	BD 2552	Steel	1	AID 1075A	
26	Screwdrivers, small, M.G., Mk 1	#	BD 2573	Assembly	1	SAID 1402B	With wooden handle
27	Screwdriver, large, M.G., Mk 1	#	BD 2574	Assembly	1	SAID 1402B	With wooden handle
28	Tool, combination, Vickers, .303-in. M.G., Mk 2	#	BD 2654	Steel	1	SAID 1679	
29	Spanner, muzzle-attachment, blank, Vickers, M.G. .303-in.		BD 2641	Steel	1	DD(E) 113/5	
30	Punch, No. 3, M.G., Mk 1	#	BD 2530	Steel	1	SAID 1403A	Hardened steel
31	Punch, No. 5, M.G., Mk 1	#	BD 2532	Steel	1	SAID 1225B	Hardened steel
32	Wallet, Vickers, .303-in. M.G., Mk I* (empty) .	#	BD 2690	Material	1	SAID 2079	Available filled, part no. BD 2691
33	Rule, slide, M.G., Mk 4	#	BD 2565	Wood	1	DD(C) 57G	Used in place of range tables
34	Tubing, condenser, steam, Vickers, .303-in., M.G. Mk 3	#	BD 2672	Brass, rubber	1	DD(E) 1208	Not illustrated

Chapter Nine

A user's view

How the Vickers performed in combat is told in the words of the men who used it across a timespan of more than sixty years, from its first use in South Africa, through two World Wars and in the Korean and later conflicts. The guns were employed in green fields, deserts, jungles and mountain terrain, working faultlessly in bitter cold, torrential rain and blazing heat.

OPPOSITE This is a photograph of the author c.1983 with his own 1917 Vickers gun, in the days when owning such weapons was possible and the cost of shooting 3,000 rounds of Second World War surplus .303 ammunition in an afternoon was not prohibitive! *(Author)*

It is perhaps fitting that the final words on the Vickers gun in service should be those of the users – those who manned the guns in war for a time period of over 60 years. The first time a Maxim saw combat by British troops was during the First Matabele War (October 1893–January 1894) and it was a portent of things to come. In the early morning of 25 October 1893, a column of 700 soldiers from the British South African Police camped in a defensive *laager* (a circular defensive position) next to the Shangani River. Unknown to them, the Matabele king, Lobengula, had moved into battle positions a force of up to 6,000 men (the number is disputed), many armed with modern Martini-Henry rifles. The British possessed several Maxim machine guns and they were immediately brought into action. In a later memoir, Major Frederick Russel Burnham, an American serving with the British force, wrote:

Then the Maxims found the range and swept the Matabele down like blades of grass before a farmer's sickle. The sun augured mercilessly on this scene of dismemberment and destruction. Unnatural noise filled the air, a sonic horror. Hyenas and vultures awaited joyfully. Three times the Matabele regrouped to launch attacks . . . finally the Maxims opened up to mow down the survivors. It was the worst military defeat that Burnham could conceive of and he was sickened at the frightful squander of human life. If the Matabele commander planned a fifth assault that day, it never came. The firing died away and in time the troopers rode out to survey the battlefield. The number of dead was enormous. At least eight hundred bodies lay in the field, some warriors dismembered by the concentrated machine gun fire. The wounded appeared to number equally high. The ultimate count was that fifteen hundred blacks died as a result of the forty-five-minute engagement. The officers called roll to assess casualties; four men killed and seven wounded.

It was an example of how technology was about to change the face of battle, although the British soldiers involved could not possibly have grasped its significance at the time, although a few others did. In the wake of the massacre, Hilaire Belloc wrote scathingly in 1898: 'Whatever happens we have got the Maxim gun, and they have not.' But it was not an easy revolution, for the fact was that the British officer class in particular were dismissive, even scornful, of the new weapons. They were cumbersome, could not be ordered about in a suitably military manner, required vast amounts of ammunition and did not in any way adhere to the strongly held ideals of a traditional, gentlemanly form of warfare. Machine gunners were simply not like proper soldiers. In his autobiography, Maxim himself records the comments made by an unidentified senior British officer.

Upon the subject [of machine guns] . . . he declared himself opposed to the idea, and listed among his objections the fact that 'the guns were not as a rule made for actual warfare, but for show' and that the gun I had designed was extremely ugly as compared with the graceful form of existing guns.

There was perhaps some excuse for this, for the guns were still regarded as an extension of the artillery, and their early use was predominantly in Colonial wars against poorly armed native warriors, where the British expected to win regardless. And frankly, no one quite knew what to do with them, but Britain soon had a taste of its own medicine during the Second Boer War (October 1899–May 1902) when they faced Boer soldiers armed with Maxims, including the 1-pounder 'Pom-Pom', an experience that all involved found extremely unpleasant, for without artillery, the British were powerless to retaliate. Indeed, Maxims became more prolific as the war dragged on, the open terrain being perfect for laying down long-range fire. Unfortunately, the Boers seldom stayed still long enough to become targets. It was not until a war erupted between Russia and Japan over Manchuria (February–September 1905) that military observers really began to sit up and take notice of the abilities of these new guns. The Russo-Japanese War was, at least in European eyes, a *proper* war, and the effectiveness of the Russians' guns could not be ignored.

LEFT The Northumberland Hussars with their two Maxims, which were privately purchased for service during the Boer War. (Author)

On January 8th, 1905 near Lin-chin-pu, the Japanese attacked a Russian redoubt armed with two Maxim guns. A Japanese company, about 200 strong, was thrown forward in skirmishing order. The Russians held their fire until the range was only three hundred yards. . . . In less than two minutes they fired about a thousand rounds, and the Japanese firing line was literally swept away.

For the first time, the international observers from Europe and the United States were able to see at first hand the defensive possibilities of the machine gun when properly deployed. It would be an exaggeration to say it forced a global reappraisal, but its effects could not be ignored.

By 1910 Germany had reassessed its strategy regarding Maxim guns – how they were to be best employed tactically and what

LEFT An interesting picture of a gun team setting up a Vickers tripod and ammunition on the ground, but also making use of a captured German MG08/15 light machine gun. The gunner nearest the camera wears the padded waistcoat issued for tripod carrying. (Author)

quantities were required – and came to the logical conclusion that they could not have enough. Britain, on the other hand, still clung to the largely imaginary Victorian ideal of romantic warfare, of flashing sabres, gallant cavalry charges and heroic soldiers clinging to the colours in the face of all odds. This may have been true in 1414, but it was palpably untrue by 1914. To an extent, the same criticisms could be levelled at the United States as well, who also prevaricated about the type and numbers of guns they needed. So when Germany began its advance across Europe in early August 1914, they were the only fighting power who had an inkling of how fundamentally important the machine gun would become.

The first British troops to meet the German advance were armed with Vickers-Maxims, two per battalion, and these were predominantly set up to provide direct frontal fire to defeat the massed waves of advancing Germans. Indeed, few realise that the first two Victoria Crosses of the war were awarded to machine gunners. On 24 August 1914, two Maxims of the machine gun section of the Royal Fusiliers were placed at one end of the vital Mons-Condé Canal Bridge at Nimy in Belgium. They held off repeated assaults by the Germans, but eventually fell silent as the crews were gradually killed and wounded. When their commander, Lt J.M. Dease, was also killed as he examined a damaged gun, Pte Sidney Godly, a trained machine gunner, ran forwards and manned one Maxim by himself, keeping up fire for two hours while the Fusiliers organised a withdrawal, despite being hit by a large shell splinter in his back and taking a bullet through his head, each of which should have been fatal wounds. Godley was clearly made of tough stuff, for the Germans were only able to overrun the bridge when, with the Fusiliers gone, his Maxim ran out of ammunition. He was captured alive and spent the rest of the war as a prisoner, being awarded his VC by King George V on 5 February 1919. Lt Dease received a posthumous VC. Seven others were to be awarded to MGC personnel during the war.

The advancing Germans were not initially aware of the British Army's lack of machine guns, and after the attacks on Langemark

BELOW One of the most familiar images of a Vickers in action, with its Sangster bipod in place and its legs strapped to the water jacket. Both men wear PH (phenate helmet) gas helmets, presumably for the benefit of the camera. The author has tentatively identified this as a gun of the 34th Battalion MGC providing covering fire at La Boisselle, 1 July 1916. *(Author)*

(21–23 October 1914) when 15 rounds of rapid fire laid down by seasoned British soldiers resulted in the destruction of two entire German regiments, German intelligence reported that the British Army were heavily equipped with machine guns. They weren't, but the few guns they did have certainly were put to effective use. Pte Sidney Holliday of the 1st King's Royal Rifle Corps was a machine gunner and recalled how, in November 1914, his two guns fought off a surprise attack:

It was just coming up dawn, we were in the trench, brewing up because it was freezing, when our sentry said very quietly 'I think the Huns are in No man's land.' We immediately stood to, Billy took the sand-bags off the gun and I cocked it and waited to see what would happen. Our Sergeant fired a Verey light and we saw a dim line of figures moving towards us. He yelled 'Fire' and I pressed the triggers on the Vickers. It opened up and I heard our other gun, about two hundred yards away also start up. I aimed low and traversed slowly. Left–right, left and the figures just melted away. It didn't seem real somehow, the noise was deafening, but all I was conscious of was that gun firing and doing what I had been so well trained to do. After we had put three belts through it, everything stopped, it was suddenly silent. Just like that. Then all we could hear was the crying of the wounded Germans. I can still hear their screams now, and I wondered then what sort of a terrible weapon I had been put in charge of. I was only 19.

It became crystal clear to the Germans that despite the paucity of British machine guns, such frontal assault tactics were both wasteful and ineffective and for the rest of the war, they rarely gave the British similar opportunities. Increasingly, their own guns were sited for defence and it was a strategy that paid off when the ill-fated mass attacks were launched on the Somme on 1 July 1916. This strategy is all the more puzzling as at Loos in September 1915 British casualties amounted to 20,000 men, mostly through machine gun fire. Yet it appeared that the British high command had

BELOW The MGC were known as the 'Suicide Squad' for good reason. Always targeted by enemy artillery and often left behind as a sacrificial rearguard, they suffered a disproportionate number of casualties compared to the infantry. This gunner at Cambrai in 1917 clearly put up a stout defence before being killed. *(Author)*

yet to comprehend the virtual impossibility of launching attacks against German positions that were heavily defended by enfilading Maxim guns. It was believed that the massive, week-long pre-attack artillery bombardment prior to the Somme Offensive would destroy the German machine gun positions, which it signally failed to do. On average a German gun crew could bring an MG08 into action from the shelter of a 30ft-deep dugout in under five minutes. Some 80% of the 57,470 British casualties inflicted of the first day of the attack on 1 July were the result of machine gun fire. A vivid first-hand account of the futility of such assaults was written by Lt Col Graham Seton-Hutchinson, whose 100th MG Company were dug in on the ridge overlooking the objective of the attack, Martinpuich.

> I . . . could see the men of the 1st Queens passing up the slope to Martinpuich. They were awkwardly lifting their legs over a low barbed wire entanglement. Some two hundred men, their Commander at their head, had been brought to a standstill at this point. A scythe seemed to cut their feet from under them and the line crumpled and fell, stricken by machine gun fire.

During the six long months of the battle, the British Vickers gun crews were mostly employed with providing long-range support fire, as the Germans very sensibly refrained from launching any assaults from the safety of their deep trenches. However, when grouped en masse to provide plunging fire, the guns gave a good account of themselves. One of the best-documented evidences of this was the role played by ten guns of Hutchinson's 100th MG Company on 24 July, when they were ordered to break up a German counter-attack behind High Wood. From their positions on the opposite Bazentin Ridge, they had to fire at a range of about 2,000yd, dropping their bullets into the trenches themselves and the invisible, sloping ground behind the wood. Preparation had to be made well in advance, and for two days ammunition and water was carried up by relays of gunners to the gun batteries, but such were the quantities of ammunition required that one man was employed continuously at a belt-filling machine for 12 hours. Although the number of rounds expended is disputed, there was no doubt that the barrage was extremely effective. As Hutchinson later noted:

> *Prisoners examined at Divisional and Corps headquarters reported that the effect of the barrage was annihilating. The strong counter-attacks . . . were broken up. The efficiency of the Vickers guns was astonishing, even to myself, well experienced in their use.*

As experience grew within the MGC, so too did the tactical demands. Crews were increasingly being sent forward with advancing infantry to engage targets that were beyond the ability of normal infantry small arms to deal with. Increasingly in the final year of the war during the Great Advance, this came to mean taking on German machine gunners protected by concrete emplacements. In the Ypres Salient, these had always been a major problem, holding up and even halting attacks, as MG bunkers lined the front in their hundreds, built in a series of staggered lines to provide interlocking fields of covering fire. They were shell-proof to all but the heaviest-calibre artillery, the firing slits provided a near-impossible target for riflemen to hit, and the only method of capture was a series of costly individual assaults that would eventually enable the infantry to work their way around them. For the Third Battle of Ypres, better known as Passchendaele, which began in July 1917, gun crews were instructed to advance with the infantry, using Sangster mounts rather than the heavy Mk IV tripods, providing them with more flexibility and faster movement, but the Sangsters sank in the mud. Bunkers were to be assaulted by infantry teams armed with smoke grenades, while the Vickers guns poured covering fire into the firing slits, preventing the Germans from retaliating. That at least was the theory, but in practice it seldom worked, as Canadian gunner Reginald Le Brun later recalled.

> *We carried the guns forward and moved off with the attack at first light. The going was terrible, we left the [duckboard] track as it was under observation and being*

shelled heavily, and staggered through the mire up to the line. Well, it wasn't a line really, just a series of water-filled holes with miserable infantry in them. The dead lay everywhere, most half submerged in that awful mud and we set the Vickers up on its tripod, using ammunition boxes packed on top of ground-sheets to try and stop it sinking; we had barely got it loaded when a shell crashed into the ground two dozen feet away, covering us all in liquid mud. Two of the crew were killed outright and my ears were singing so badly I couldn't hear a thing. The gun was blown into the mud and the belt was soaked. We tried to clean it with water from the tins but we then realised that the other boxes of ammunition had vanished. We had a gun, but nothing to fire in it and we had to get more brought up, which took forever. We were stuck there helpless and shell after shell landed around us. By evening I was the sole survivor of the team and when we were relieved I carried the gun back on my own.

Amazingly, during that time a Canadian official photographer crawled up to the lines, and took a picture of Reg and his crew, forlornly looking at the lens, which has become one of the most iconic images of Passchendaele. Eventually the fighting broke out of the stalemate of the trenches, and by 1918 a war of movement had begun. In conversation with the wonderfully named Lt Col Julius Caesar (the name was a proud family tradition) in the late 1970s, the author asked about the role of the MGC during this period. His response was informative:

> Well, in fact, by this date the Lewis gun had really replaced the Vickers as the primary front-line machine gun. After all, it was lighter, needed only two men to man it and of course, it had a more than adequate range for most situations. In our [machine gun] company, we actually had a couple that we had scrounged, as sometimes they were more useful. We didn't tend to use the Vickers for long-range fire so much by then, it was mostly moving with the infantry and

ABOVE A less common image of the same gun team. From the setting of the rear sight, they would appear to be firing at a range of between 2,200 and 2,500yd. Note the spoons tucked into their puttees, the handiest place for easy access when food appeared. *(IWM Q003996)*

BELOW A nice image of MGC infantry in action manning a hastily prepared trench during the German Offensive of spring 1918.
(IWM Q006279)

suppressing fire against German machine gun positions or snipers. I remember at one point it did come in handy though, we were somewhere near Arras and came across a German seventy-seven field battery, who were firing shrapnel over open sights and creating havoc for the infantry. The Lewis was hopeless, I recall it was over 1,000 yards away and for the first time in ages we actually set the [Vickers] guns up on their tripods and engaged them. We saw the gunners dropping and several took cover behind their shields. It allowed the infantry to move up and they mopped them up. I didn't see any prisoners either.

One MGC veteran, Pte Thomas Williamson, recounted how, during the advance on Arras in the summer of 1918, his infantry platoon was held up by a concealed German sniper. He was asked to bring his gun up and deal with the rifleman. Using the Sangster mount, he set up the gun at 200yd, while his corporal watched through binoculars as the infantry drew the sniper's fire. The location, a small bush within a copse of trees, was identified, and the

ABOVE Foot-slogging, something all infantrymen were familiar with. This group of MGC men carry their guns and ammunition to the next position during the great advance of summer 1918. Because of the fast pace of movement, they are using Sangster mounts and not tripods. *(Author)*

gunner fired two shots, at which point a cheer rose from the British soldiers who had seen the muzzle of the German's rifle jerk into the air, then fall. A patrol found that one of the bullets from the Vickers had struck the sniper squarely between the eyes, and Pte Williamson was heartily congratulated for the accuracy of his shooting, eventually being awarded the Military Medal. He never told anyone that the only reason he fired two shots was not due to his confidence in the accuracy of his shooting, but because his Vickers gun had jammed.

By the end of the war, there were three times the number of Lewis guns in infantry service compared to the Vickers, and the machine gun wastage rate due to enemy action in 1918 was the highest of the whole war. This inevitably placed a strain on the manufacturing capabilities of the Erith and Crayford factories, which were already struggling to keep pace with demand. The Armistice in November solved the problem, as it resulted in an abrupt cessation of manufacturing. Part-assembled guns were finished and stored and much of the workforce was laid off.

Little did anyone suspect that in under 20 years these guns would once again be in action. The training course was now of seven weeks' duration, and followed pretty much the same lines as in the previous war. When the German Army swept through the Low Countries and France on 10 May 1940, there were striking similarities with 1914. The outnumbered British Expeditionary Force and French troops facing them put up a brave, but hopeless defence against the overwhelming might of the German armoured columns. Although the excellent Bren light machine gun had been introduced in 1936 to replace the ageing Lewis gun, it had a limited magazine capacity of 30 rounds and a maximum range of 600yd, so could not compete with the Vickers in terms of sheer firepower or range, although it was very widely issued and proved very successful. The decision to retain the Vickers in the hands of five dedicated machine gun regiments has already been explained, and they saw immediate action as the BEF retreated towards the French coast in June 1940. This account was given by the author's uncle, former sergeant Bill Cooke, of the Highland Division.

> We were in a small village, Warhem I think, with two [Vickers] guns and a platoon or so of mixed infantry. We were ordered to put the guns on the flanks to cover the roads, so we dug in to a small copse of trees and made the best of it. When the German tanks arrived, there was no point in opening fire, it would just have wasted ammo and given us

ABOVE Two British Expeditionary Force Vickers guns set up rather conspicuously in a French town during the retreat to Dunkirk, 1940. French officers are conferring in the background. *(Author)*

away, so we waited. Eventually their infantry arrived in several trucks, and we all let rip. It was only about 400 yards and they went down like skittles. Then two tanks came up and began raking the trees with gunfire, they had twin machine guns. We couldn't really do anything, then another [tank] arrived with a big gun and systematically began destroying the buildings. We packed our guns and tripods on our shoulders and legged it back through the village as fast as we could. We were pretty well out of ammunition anyway and we were later told to abandon them as we retreated, which broke our hearts. I recall taking the lock from mine, and we put them on the road and let the trucks drive over them. I brought that lock home with me. [The author has that lock now, and it sits on his desk.]

In other theatres of war, the Vickers proved its worth once again, particularly during the fighting in the Western Desert, where there were vast open expanses that could be very effectively covered by machine guns that, once dug in, were almost impossible to locate. The determination of Vickers crews in action was exemplified by the action of Capt James Jackman of the 1st Battalion Royal Northumberland Fusiliers south of Tobruk on 21 November 1941. He led his machine gun company in 15cwt trucks on to a ridge in an attempt to knock out their anti-tank guns which had stopped the British tank advance. The site was being swept by intense German gunfire. As his citation later described:

> He immediately started to get his guns into action as calmly as though he were on maneuvers [sic] and so secured the right flank. Then, standing up in the front of his truck, with calm determination he led his trucks across the front between the tanks and the German guns – there was no other road to get them into action on the left flank. Having forced the German gun crews into cover, the advance began again, and he was subsequently awarded the Victoria Cross.

ABOVE Derna, North Africa, January 1941. The gunners have wrapped their water tin in sacking to try to reduce the signature plume of steam that always escaped and could give away the position of a gun to the enemy. *(IWM E1819)*

Static long-range firing was highly effective during the invasion of Italy in 1944. Vickers guns were heavily employed to provide covering fire for assaults on German positions. Bill Cooke landed in Sicily in July 1943 and was immediately employed with his Vickers team in providing supporting fire for the attack on Pachino, which enabled the Canadian troops to advance.

RIGHT A camouflage-netted gun fires in support during the battle for Tobruk in 1941. Ammunition tins have been readied. Curiously, while the tins and tripod have been painted in desert khaki, the gun has not. *(Author)*

RIGHT Vickers gunners of the 2nd SAS at Castino, Italy, in spring 1944. The man on the left has the tripod front legs over his shoulders, the most comfortable method of carrying it. Their mix of British and American sidearms and casual appearance could easily lead to them being mistaken for local partisans. (IWM NA25407)

BELOW This snapshot gives some idea of the dense jungle in New Guinea that made fighting such tough work for the Vickers crews. It could rarely be used efficiently in such terrain, and the Bren gun became the preferred light machine gun. (Australian War Museum 063095)

We set up three guns at dawn and fired at about 1,500 yards, putting tracer into the buildings, some of which caught fire. We must have fired twenty or thirty thousand rounds and we learned afterwards that the buildings looked like colanders. I don't think any Italians stayed to face us, they [the Canadians] said they surrendered wholesale and were pleased to get away from the 'mitragliatrici terribili' [terrible machine guns]. We got a mention [in dispatches] for that.

Of course, the Vickers turned up in places that one would not normally associate as being the perfect environment for their use. Possibly the most improbable was in the jungles of Burma, where Chindit units carried their Vickers through swamps, dense jungle and over successions of steep ridges, in appalling heat and humidity. Because of the permanent damp, the guns had to be constantly kept oiled and inspected daily. One veteran recalled that the barrels and working parts would rust internally in less than an hour if not immediately oiled. Belts had to be kept in sealed tins, and only opened when required, and even then the cartridges often had to be cleaned of the verdigris that formed. Nothing was carried that was not of absolute necessity. Mules were normally employed, the guns being broken down into portable parts, as indeed they had been in the First World War. Supplies of ammunition had to be carefully conserved due to the difficulty of resupply in the jungle and it was rare that the guns could be fired at anything other than relatively close quarters, the terrain usually precluding long-range fire. Predictably, despite the affection that the gunners had for their guns, lugging one through the jungle was not considered much fun. An unnamed soldier from the King's Liverpool Regiment, which had kept their Vickers because of their Chindit role, later recalled:

The Vickers was a wonderful gun and we were proud to be gunners, but my God, they were bloody awful things to carry. We had to leave the mules behind sometimes and after a few hundred yards they [the guns] felt like they weighed a ton. You often tripped up on the submerged roots of the trees, and if you dropped the gun it needed cleaning of course, though we sometimes just set them up, wiped them over and fired and they seldom went wrong, I think they were indestructible.

In the fast-moving war that began after the Allied invasion of France on 6 June 1944, the Parachute Regiment were issued with them, which posed some problems when dropping paratroops, who clearly could not carry a heavy gun or tripod without risking injury. They were loaded into special drop containers and a good

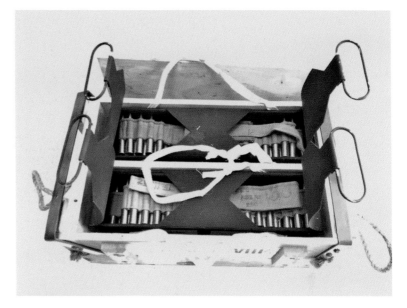

ABOVE A pair of post-Second World War sealed ammunition tins with stripless belts of Mk VIII ammunition in their wooden transit case. *(Richard Fisher)*

LEFT A Vickers section of the 2nd Middlesex Regiment waiting to open a barrage over the Maas–Schelde Canal, prior to the attack launched on 20 September 1944. Double-parabolic flash hiders and dial sights are fitted to all guns. The gunners wear the new 1944-pattern steel helmets. *(IWM B10144)*

LEFT A suspiciously clean Vickers team, carrying their gun and tripod. The men wear British Winter Combat 1952-pattern fur-lined windproof jackets with attached white gloves; the second gunner is also wearing the trousers. The last man carrying the ammunition also had an Enfield No 4 rifle, unusual for a Vickers team. *(IWM KOR000699)*

crew could empty one, set up a Vickers and be able to fire in under five minutes, always assuming, of course, the containers could be located. During Operation Varsity on 25 March 1945, the Vickers teams found themselves on the drop zone on the edge of the objective, a town called Hamminkeln, but missing their guns. Cpl Barry Lees later recalled:

> *When we landed we couldn't find the containers so we launched the attack without our guns and managed to secure the town. Eventually someone reported seeing some parachutes in the Issel [river] so we had to get boats and fish them out, though we lost a couple, which had broken free. It took hours to clean everything and the belts were all useless. We had no loading machine so we all had to hand-load the new belts and it took bloody ages.*

Some idea of the workload of the Vickers gunners can be gleaned from the fact that the Cheshire Regiment recorded their expenditure of .303in Vickers ammunition for the period of September 1943 to April 1944 while fighting in Italy, where they fired a little under 5.5 million rounds. One MG company of the Kensingtons alone fired 500,000 rounds (burning out 34 barrels in the process) after landing in Normandy. In fact, .303in ammunition expenditure during the 1944 advance across France, Belgium and the Netherlands was as great as that of the First World War, showing that the tactical usage of the Vickers had by no means diminished. By the end of the war, the Vickers MG regiments had fought through every theatre of war and in every type of climate. But mortars and heavy machine guns, such as the Browning M2, increasingly took on the role of long-range support weapons.

BELOW A nice image showing the perfect terrain for using the Vickers in Korea. The field of fire from their position is almost unlimited. *(Author)*

Besides, by 1945 transporting a Vickers battalion of 36 guns (reduced from 48 but by then replaced by 4.2in mortars) had become a major logistical issue, requiring 99 15cwt trucks, 13 3-tonners, one 30cwt lorry, 9 cars and anything up to 50 despatch riders. The Army was looking closely at the necessity of employing such huge resources and took the decision to disband the MG battalions, supplying instead six guns to each infantry battalion. The Machine Gun Training Centres in Cheshire and the school at Netheravon were closed. However, fate in the form of the Korean War intervened in 1950 and Vickers teams from Britain and Australia saw action.

Though all of the guns employed in Korea were of Second World War vintage, there was little difference in equipment or tactics from the teams that had gone into action on the Western Front in 1916. Two things set the Korean War apart, though: the first was the need to fight in bitter winter weather and, second, the terrain. The cold was a major factor. Lined trousers, hooded parkas and mittens were issued to Vickers crews, who often had to sit in static positions for long periods. The other factor was that the long ranges at which the gunners most frequently fired kept them safe from all but occasional enemy artillery fire, although getting supplies to and from their high, remote positions was hard work, as ex-sergeant Dave Laddis of the MG section, Durham Light Infantry, commented:

> *We were quite high up, about fifteen hundred feet, and the only route was a narrow rocky path which was slippery with snow and ice. We were well dug in with rocks we built around us and a couple of tarpaulins we had lugged up because the wind was fearsome and cut you like a knife, but anything we forgot had to be fetched and it was an hour to get down and back up again. We had a rota of who was next to go out and no-one ever wanted to go of course, so we conserved our ammo as much as possible. But when we were firing in support of an attack we constantly had blokes up and down that damn track with ammo and water. The one saving grace was we could brew up when we wanted as*

ABOVE An Australian gunner of the 3rd Royal Australian Regiment fires in support of an assault on Pakchon, Korea, in October 1950. Note the ammunition tins opened in readiness and the Owen sub-machine gun next to him. *(Australian War Memorial HOBJ1658)*

> *we were over three thousand yards from the Chinks [Chinese Nationalist Forces] so the kettle was permanently on, but you had to make sure the tins were only filled with water and not coolant!*

By this time it was clear that the British Regular Army was going to have to become a leaner, more efficient force, and this didn't include making use of weapons designed in the Victorian era. It was not what a modern, professional army wanted. In fact, the concept of a multi-role machine gun had radically influenced all post-Second World War Allied military thinking and Britain was no exception, having been heavily influenced by exposure to the Germans' general-purpose MG34 and MG42 machine guns. So Korea proved to be the swansong for the Vickers in a major combat role. It lingered on, providing limited support in peacekeeping roles in Africa, Borneo, Malaya and the Middle East during the late 1950s and early 1960s. The Parachute Regiment hold the laurels for the final recorded action of the Vickers, in Aden in 1968, by which time it had been replaced by the 7.62mm L7A1 General Purpose Machine Gun, adopted in 1962. The grand old lady had finally been laid to rest.

In retrospect

Despite the genuine belief of men such as Gatling and Maxim that their inventions would simply make warfare too horrendously costly to contemplate (later mirrored in the technology behind the atomic bomb), in reality there was no doubt that the Vickers, like all weapons, was inevitably destined to become nothing more than an extremely efficient machine for killing. Science did not move at the pace it does today, and so a complex, expensive gun such as the Vickers-Maxim was considered to be the apogee of weapons technology. The Victorians, with their unshakeable self-belief, felt that nothing more efficient was ever likely to replace it. So the guns adopted into service were built to last with a level of craftsmanship that was eventually to prove economically unsupportable. This is reflected in the many hundreds in the hands of shooters and collectors that are still working today. But as we are only too well aware now, nothing speeds up weapons development faster than a war and in the wake of two world wars, the Vickers eventually lost its crown as the queen of the battlefield, being relegated to a supporting role as a medium machine gun, secondary to other more practical and cheaper weapons. In the late 20th century, as armies became increasingly run by government accountants, so the requirement for mass-produced, cheaper weapons became the norm. Wars aren't won by giving soldiers beautifully made guns, and the Vickers concept of quality and longevity was no longer relevant, nor sustainable.

So, did the Vickers guns that had been used so effectively on the battlefields of the First World War actually change the course of history? Had they not existed, then the likelihood of a German victory in 1914 based simply on overwhelming odds would have been that much greater. It could be argued that the

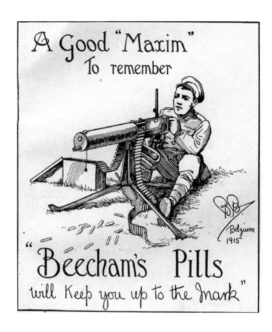

LEFT The Vickers was even used in advertising, in this case drawn by Bruce Bairnsfather for Beecham's Pills, though they do not specify for what purpose; possibly to clear a stoppage! *(Lawrence Brown)*

OPPOSITE A small piece of Vickers history. The lock taken from the Vickers gun abandoned by Sergeant Bill Cooke during the retreat to Dunkirk in 1940. He carried it in his pocket all the way back to England and used it later in the war when instructing on the Vickers guns. It lived in pride of place on his mantelpiece for as long as the author could remember. *(Author)*

LEFT One of the last photographs of Hiram Maxim taken prior to his death in November 1916. Bedecked with honours and hugely wealthy, he never made public his thoughts about the slaughter wreaked by his guns during the First World War. *(Author)*

LEFT The famous British cartoonist Bruce Bairnsfather used the Maxim in several cartoons, this being the most well known, dating from early 1915. *(Lawrence Brown)*

RIGHT During the First World War the Vickers gun had become an iconic weapon and a huge range of Machine Gun Corps souvenirs bearing its image were produced. Much was home-produced trench art, but items like this ashtray were commercially manufactured. *(Author)*

First World War served only to set the rules for the Second, but it is a truism that both conflicts changed forever the balance of global power, and in doing so altered the world map. This heralded a whole new socio-political order, with greater freedoms than anyone could have possibly imagined in 1914. Women worked, earned and had careers and a decade after the war, British women got the vote, as did American and most other European women, and by 1948 Britain had ceded most of its colonies. But the First World War had also promoted the advance of mass-production techniques to a whole new level, making it possible to manufacture more cheaply and in greater quantities than ever, as can be glimpsed from the production figures of the Vickers in 1918. This, of course, is something we accept as perfectly commonplace today, in a world where mass production and globalisation are now an accepted part of our lives.

However, in the late 19th century, no single country possessed the ability to provide its armies with what would now be termed weapons of mass destruction and Maxim can be forgiven for thinking he may actually have solved the problem of war. He died on 24 November 1916, just as the bloody Somme battles fizzled out in the mud and rain of winter. What he thought of the use to which his guns had been put he never made public, but one suspects he must have felt both sadness and bewilderment. Alas, mute testimony to the impartial efficiency of his invention can be seen in the tens of thousands of headstones scattered today over the old battlefields of the First World War and later conflicts. There was a reason that the Machine Gun Corps chose as their memorial figure the biblical 'Boy David', who killed Goliath. On the statue erected to them at Hyde Park Corner is a fitting comment from the Bible to the terrible legacy that Maxim left to the world:

'Saul hath slain his thousands, but David his tens of thousands.'

LEFT The Vickers gun has featured in at least 68 films. This poster is from one of the more unlikely, the 1965 film *Viva Maria!*, starring Brigitte Bardot and Jeanne Moreau. Amazingly it won a BAFTA. *(Author)*

LEFT 'The Boy David' on the Machine Gun Corps Memorial at Hyde Park Corner, London. Designed by Francis Derwent Wood, it was unveiled on 10 May 1925. It commemorates the 51,049 officers and men who were killed, wounded and missing between 1915 and 1918. The popular story that the two Vickers guns on the plinth are real and coated in bronze is sadly a myth. *(Author)*

Bibliography and selected reading

Chappell, M., *The Vickers Machine Gun* (Hatherleigh, Wessex Military Publishing, 1989)

Chinn, G.M., *The Machine Gun: History, Evolution and Development*, 3 vols. (Washington, Department of Defense, Bureau of Ordnance, 1951)

Coppard, G., *With a Machine Gun to Cambrai* (London, Imperial War Museum, 1980)

Crutchley, C.E. (ed.), *Machine Gunner, 1914–1918* (London, Purnell Books, 1975)

Ellis, J., *The Social History of the Machine Gun* (London, Croome-Helm Ltd, 1975)

Goldsmith, D., *The Devil's Paintbrush: Sir Hiram Maxim's Gun* (London, Greenhill Books, 1989)

Goldsmith, D., *The Grand Old Lady of No-Man's Land: The Vickers Machine Gun* (Cobourg, Collector Grade Publications, 1994)

Handbook, *.303 Vickers Machine Gun and Tripod Mounting, Mark IV.* (Kingsway, London, HMSO, 1939)

Hawkey, A., *The Amazing Hiram Maxim* (Staplehurst, Spellmount Ltd, 2001)

Hutchinson, Lt Col G.S., *Machine Guns: Their History and Tactical Deployment* (London, Macmillan, 1938)

Hutchinson, Lt Col G.S., *Warrior* (London, Hutchinson and Co., 1932)

Kay, R., *27th Machine Gun Battalion: New Zealand Official History of World War Two* (Wellington, Whitcombe and Tombs, 1957)

Lloyd George, D., *War Memoirs of David Lloyd George, 1915–1916* (New York, Little, Brown and Company, 1933)

Maxim, Hiram, *My Life* (London, Methuen and Co., 1915)

McLean, D.B., *Handbook of the US Vickers Machine Gun, Model of 1915* (Arizona, Normount Publishing, 1973)

Scott, J.D., *Vickers: A History* (London, Weidenfeld and Nicolson, 1963)

Segel, R.G. (2012), 'US Colt Vickers Model of 1915', *Small Arms Defense Journal*, 20 February 2015. Available at: https://www.smallarmsreview.com/display.article.cfm?idarticles=2940

Skennerton, I., *.303 Vickers Medium Machine Gun* (Labrador, Australia, Skennerton Publishing, 2005)

Temple, B.A., *Identification Manual on the .303 inch British Service Cartridge, Vols 1–4* (Burbank, Australia, Temple Publishing, 1986)

The War Office (1941), *Small Arms Training Manual, Vol. 1, Pamphlet 7. Parts I, II and III. The .303 inch Machine Gun* (London, HMSO, 1941).

A full-length view of the rare air service Colt-Vickers chambered for the 11mm Desvignes cartridge. Much lightened, with an external trigger mechanism and steel-linked ammunition, it provided both reliable service and a high rate of fire.
(Courtesy International Military Antiques, ima-usa.com)

Appendix

Vickers gun specifications

Gun, Machine, .303in Vickers Mk I
Weight of gun (only): *32.5lb*
Weight of tripod, Mk IV: *50lb*
Method of cooling: *Water*
Weight of water (7½ pints): *7.5lb*
Length of barrel: *28¼in*
Length of gun: *45½in*
Life of barrel (approx.): *20,000 rounds*
Accuracy of barrel (approx.): *5,000 rounds*
Recoil friction: *Not exceeding 4lb*
Recoil spring tension w/o muzzle booster: *Up to 8lb*
Recoil spring tension with muzzle booster:
    Between 8 and 10lb
Life of main spring (approx.): *6,000 rounds*
Rear sight graduated to: *2,800m*
Life of barrel packing: *5,000 rounds*
Ammunition: *.303in Mk VII ball. Later Mk VIIIz*
Weight of a loaded ammunition box: *21lb*
Weight of a loaded belt: *15¼lb*
Weight of bullet: *174 grains*
Weight of powder: *47± 1½ grains cordite*
Chamber pressure: *50,000psi*
Muzzle velocity (approx.): *2,500fps*
Belt capacity: *250*
Average rate of fire: *500r/pm.*

Gun, Machine, Vickers 0.5in Mk III
Weight of gun empty: *56lb*
Weight of gun filled: *66lb*
Length of gun: *52in*
Method of cooling: *Water*
Calibre: *0.5in*
Bullet weight: *580 grains*
Muzzle velocity: *2,500fps*
Rate of fire: *500–600r/pm.*

Vickers High-Velocity Automatic Gun. 12.7mm
Weight of gun empty: *101lb*
Weight of gun filled: *122lb*
Length of gun: *73in*
Method of cooling: *Water*
Calibre: *12.7mm*
Bullet weight: *690 grains*
Muzzle velocity: *3,000fps*
Rate of fire: *350–400r/pm.*

OPPOSITE **The training of machine gunners was overhauled in the late 1930s to make the courses both shorter in length and broader in content. Here a training circle of guns has been formed, so that the instructor can be simultaneously seen and heard by all the gun crews.** *(IWM H011861)*

Index

Accessories 59, 61
  aiming lamp 61
  ancillary equipment 145
  armoured nose cone 95
  barrel wear gauge 103
  cheek-pads 92
  improvements 71
  parts kit 64
Accles, James 20
Aden 97, 161
Advertisements 15, 59, 107, 163
Aerators factory 74
Aerial warfare 89-93
  dogfights 91
Afghanistan 117
Agar (Ager), Wilson 16
Agar/Ager guns 9, 18
  Union Repeating Gun 16
Aircraft
  Albatros D1 91
  Bristol Fighter 91
  Nieuport 28 92
  RE8 81, 92
  Sopwith Dolphin 92
  Sopwith Snipe 93
Aircraft armaments 9, 89-90, 99, 107, 111, 131, 167
  synchronisation/interrupter gear 89-90
    Constantinescu system 90, 92
    manufacturers 89-90
    Fokker-Lubbe system 89
A Life of Pleasure play, Princess Theatre, London 9
American Civil War 14, 16-18
Ammunition – see also Calibre and Cartridges 39, 58, 61, 71, 73-77
  checking 129, 131, 129
  centrefire 21
    American Berdan 21
    British Boxer 21, 30
  ejected from gun 97
  metallic 15, 18
  Mk VI 75
  Mk VII 51-52, 74-75, 77
    range table 75
  Mk VIII/VIIIz 73-74, 76-77, 117, 159
  quantities used 85, 112, 148, 152, 158, 160
  rimfire 18, 20
  Vickers-Maxim 20
  .45 Gardner-Gatling 20
  .45 Maxim Mk. VII 20
  .450in Martini-Henry 20, 30
Ammunition belts and drums 32-33,39, 49, 51, 56, 58, 61-63, 71-74, 89-90, 93, 94-95, 97 , 102-103, 112, 129
  cloth/webbing 90-91
  filling by hand 74, 91, 131, 133, 160
  filling machines 61, 92-93, 131, 133, 152
  inspecting 131
  Mk II 90
  Mk III 90
  Prideaux pattern 90, 93
  sealed tins 158-159
  steel-linked 90-91
  stripless 72-73, 159

Ammunition boxes and containers 33, 61-64, 72-74, 81, 83, 86, 89, 93, 131, 159
  earth-filled 71
  metal 63-64, 72
Ammunition production 40
  production figures 73, 93
Anti-aircraft guns 71, 75, 81, 87, 102, 108, 110
Armoured Fighting Vehicles (AFVs) 94, 100-103, 111
  Carden-Lloyd Tankette 101
  M2 Light Tank 101
  M3 Stuart 101
  Seabrook lorry 94
  Universal Carrier 101, 115
  Vickers Cruiser Mk I 101
  Vickers Medium Mk II 10
Australian Army
  3rd Royal Australian Regiments 161
Australian Imperial Forces 108-109
Australian Vickers guns – see Lithgow-Vickers

Bairnsfather, Bruce cartoons 163-164
Baker-Carr, Brig Gen Christopher D'Arcy Bloomfield Saltern 80
Bardot, Brigitte 165
Barnes, Charles E. 16
Barrels 14, 66, 70, 74, 110
  changing 65, 97, 121, 123, 137, 139, 160
  cleaning 121
  Mk I 55
  multiple 16-18, 27
  rifling 14, 16, 18, 65
  rotating 14, 16
  smooth-bore 14
  twin 17, 27
Battery guns 15-16
Buckham George 44
Belloc, Hillaire 148
Benét-Merciés M1909 gun 104
Berthollet, Claude-Louis 14
Billinghurst, William 16
Blacker, Maj L.V. 90
Black powder (gunpowder) 7-8, 12, 20-21, 32, 39
Boer Wars 42-43, 52, 148-149
Bravery awards 108, 150, 155-156
Breech 14, 21, 27
  obturation /windage 14, 18
  screw system 13
Breech-loading 13-15
Bren gun 155, 158
British Army 15, 18, 39-41, 51-52, 54, 70, 75, 150-151
  Argyll and Sutherland Highlanders 87, 95
  Brigades Machine Gun Companies 82
  Cavalry Machine Gun Squadrons 81, 95
  Cameronians (Scottish Rifles) 87
  Cheshire Regiment 95, 160
  2nd Coldstream Guards 80
  Colonial and Dominion units 107-108
    Australian 108, 110-111
    New Zealand 108-109, 111

Devonshire Regiment 95
Durham Light Infantry 161
Gordon Highlanders 52
Guards Machine Gun Regiment 82
Highland Division 155
Infantry Machine Gun Companies 81-82, 95, 154
  Guards Division 82
Kensington Regiment (Princess Louise's) 95, 160
King Edward's Horse 52
King's Liverpool Regiment 158
King's Royal Rifle Corps 8, 151
last new Vickers gun 96
London Divisions 86
Long Range Desert Group 94, 103
Machine Gun Battalions 82, 111, 161
Machine Gun Corps (MGC) 7, 59, 70, 81, 96, 150-153, 164-165
  Comforts Fund 59
  disbanded 87, 95
  formation 80
  memorial statue, Hyde Park Corner 164-165
Manchester Regiment 95
Middlesex Regiment 95
2nd Middlesex Regiment 71, 159
Motor Machine Gun Service (MMGS) 82, 93, 113
  Heavy Branch (later Tank Corps/ Royal Tank Regiment) 82-83, 94-95
Motor Service 82
Northumberland Hussars 149
Parachute Regiment 159, 161
1st Queen's 152
reduction in size 87
1st Rifle Brigade 84
Royal Field Artillery 82
Royal Fusiliers 80, 150
Royal Norfolk Regiment 81
Royal Northumberland Fusiliers 156
Royal Northumberland Regiment 95
Royal Tank Corps 95
Seaforth Highlanders 52
Support Weapons Wing 96-97
1st Surrey Rifles 96
Territorial units 96, 107
Weedon storage depot 97
3rd Army 86
British Board of Ordnance 13
British South African Police 148
Brooks MM, DCM, Capt J.W. 84
Brownlow, Lord 83
Browning, John Moses 41
Browning, Matthew 41
Browning 116
  M2 gun 97, 101-102, 160
  M1917 107
Burnham, Maj Frederick Russel 148
Butler, Capt J.S. 104, 106

Caesar, Lt Col Julius H. 153
Calibres (bullets, guns and ammunition) 39
  large 39
  M1930 Type D 117
  rifle 39-40, 104
  Russian 7.62 x 54mm 117

.22 8, 15
.30 109; Krag 21
.30-06 96, 106
.303 18, 39, 41-42, 49, 51-52, 62, 72-75, 86, 99-101, 106, 160
.45 17-18, 20, 30, 32, 41
.45 Martini-Henry 18, 20, 30
.45/70 20
.50 20, 97, 101-102
.577/450 62
.58 16-18, 20
8mm bottleneck 21
2.7mm 102
.5in 102
1in 20
11mm Desvignes 106-107, 167
37mm 18, 36
Canadian Saskatoon Light Infantry MG Section 96
Canadian Vickers guns – see Colt-Vickers Model 1914
Cap badges 7, 83
Carriages 20-21, 36, 38, 41, 49, 117
Cars 82, 113-114, 161
  Ford Model T 82
  Lanchester 95
  Rolls-Royce 82, 93
Cartridges – see also Ammunition and Calibre 14, 73-77
  armour-piercing 74-75
    SMK 75
  ball-tracer 74-75
  black powder 39-40
  boat-tailed 74
  brass 14-15
  capped steel 18
  linen16
  metallic 16, 24
  NATO 7.62 x 51mm 97
  paper 16-17
  self-contained 15, 30
  smokeless 39
  tracer 75
  velocity 74
Chinese Nationalist Forces (Chinks) 161
Chlorine gas 63
Clark, Sir Andrew 33
Colt Patent Firearms Mfg Co. 18, 54, 58
  Gas Hammer machine gun 41
  M1895 104, 112
  M1896 112
Colt-Vickers guns 92, 105-106, 117
  air service gun 11mm 106-107, 167
  M1914 112
  M1915 96, 105, 107
  Russian order 107
Cold War 96
Components, Maxim/Vickers guns – see also Accessories, and Water jacket
  ammunition supply to the gun 31, 91, 94, 121
  breech mechanisms 27, 33
  changes made 70
  connecting rod 123, 137

cranking handle 45, 48, 119, 135, 139, 141
  positions 141, 143
extractor sliding cartridge carrier 32-33
feedblocks 44, 49, 92-93, 121, 127, 135, 143
firing mechanism 33
firing regulator 28
fusee, spring and cover 43, 90, 106, 119, 121, 123, 125, 137
locks 43-44, 48, 55-56, 67, 70, 117, 125
  cycle 143
  Scent-Lock 14
  working positions 137
muzzle attachments 131, 159
  booster 44, 90, 100, 102, 106
  cups 58, 90, 125, 139
  deflectors 71
recoiling mechanism 30
rotary feed drum 27, 29
rotating crank 27
sideplates 54, 70
toggle mechanism 48
traversing and elevation mechanism 28, 37, 60, 85, 143
  automatic 40
trigger 60, 90, 135, 139-142
  pistol-grip 100
Constantinescu, George 90
Cooke, Sgt Bill 155, 157
Cowper, 2Lt. W. 96
Crews 81, 96, 149-150, 152, 155, 158
  aircrew 131
  Australian teams 110, 112, 161
  belt-loader 37
  Canadian 112-114
  carrying guns and ammunition 112, 155, 158, 160
  casualties 82, 150, 165
  escaped from occupied Europe 5
  German 152
  gun commander 84, 135
  New Zealand 108-109, 111-112
  No 1 gunner ('Emma Gee') 81-84, 86, 95, 108, 133, 137, 139, 148-151, 158, 160
  No 2 gun and mounts carrier 81, 133, 137
  Nos 3 and 4 ammunition carriers 81, 133
  No 5 scout 81, 86
  No 6 rangetaker 81, 96
  protective clothing 65, 82, 160-161
    facemask 82
    gas helmets 153
    gloves 65, 82
    steel helmets, 1944 159
    waistcoat 82, 149
  Russian 149
  shelter 133
Crowell, Benedict 107

Dawson, Trevor 44
Dease, Lt J.M. 150
de Courcy Prideaux, William 91
Degtyaryov DP-27 machine gun 117
Disposal of guns 70, 97, 104, 115
  Australian 111
  Canadian 116
  collectors' market 97, 111, 116, 163
  dumped in sea 97
  stored 97, 111, 116
Durs Egg, London 13
DWM (German) Maxim MG08 40, 44-45, 48, 52, 56, 58, 75, 87, 102, 149, 152

Edison, Thomas 25-26
Elisha Collier, Boston 12
Enfield Arsenal 27
Enfield guns 41, 48, 59
Enfield-Maxim 51
Engineer, The magazine 16, 19, 28-29
Films featuring Vickers guns 165
  Viva Maria! 165
Finnish Maxim M09/21 117
Firing
  accuracy 143
  at night 61, 85
  beaten zone 86
  cocking the gun 89-91, 135
  cook-offs 20, 104
  cycle 127, 137
  direct (enfilade) fire 84
  for the first time 135
  firing grip 135
  grazing fire 85
  indirect fire 85-86, 96
  jammed guns (stoppages) 91, 141, 143
  positioning the gun 85-86, 133, 156
    camouflage 157
    given away by escaping steam 157
  safety checks 12
  setting up 121, 133, 149, 153, 156, 160
  sound 135
First Matabele War 1893-94 148
First World War 17, 52, 73, 75-76, 82, 91, 96, 99, 104-117, 160, 163
  Africa 93
  Allied Offensive (Great Advance) 1918 107, 152, 155
  American Expeditionary Force 107
  Amiens 87, 103
  Armistice 155
  Arras 154
  Battle of Komarów 117
  Battle of Loos 86, 151
  Battle of Passchendaele 114, 152-153
  Battle of Tannenberg 117
  Bois Grenier 87
  Cambrai 86, 151
  Camiers Overseas Base Depot 81
  Canadian Expeditionary Force 112-116
    Canadian Machine Gun Corps (CMGC) 114-115
  casualties, British 151-152
  ceasefire 70
  cost to Britain 104
  Egypt 93, 111
  Flanders 63, 114
  flooded trenches 63
  Gallipoli 59-60, 107, 110-111
  German Offensive 1918 154
  Hindenburg Line 87
  Homs, Syria 108
  Hundred Days Offensive 1918 87, 115
  La Boisselle 153
  Langemark 150
  machine gun shortage, Allied 53 150-151
  Martinpuich 152
  mass production 164
  Middle East 93, 108
  Mons-Conde Canal Bridge, Nimy 150
  New Zealand Expeditionary Force 111
  Palestine 93, 112-113
  Russia 93
  Siegfried Line 71
  Somme 48-49, 77, 82, 94-95, 110, 114, 151-152, 164
  South Pacific 109
  USA enters the war 106
  Vickers guns sold off 70
  Vimy Ridge 114
  War bonds 107

Western Front 54, 65, 85, 93, 108, 112, 161
Ypres 66, 83, 87, 152
Flintlocks 7-8, 12-14
Flobert, Louis-Nicolas 8
  .22 lead ball 8
  6mm lead bullet 15
FN MAG L7 General Purpose Machine Gun 97, 111-112, 161
Forsyth, Rev Alexander John 8, 14
Fosbery VC, Maj G.V. 17
Franco-Prussian War 1870 9
French Army 21
Fulminate of mercury 8, 14-15

Gardner guns 17, 19, 27-28, 30, 37
  M1879 17
Gardner, William 17-18
Gatling Co. 36
Gatling guns 12, 19, 25, 28, 30, 37, 135
  electric-powered 9, 20
  hand-cranked 9, 135
  ten-barrelled 18
  M1862 20
Gatling, Dr Richard Jordan 9, 18, 28-29, 31, 163
Geddes, Sir Eric 53
German machine guns MG34 and MG42 117, 161
Godly, Pte Sidney 150
Goldsmith, Dolf *The Grand Old Lady of No Man's Land* book 55
Great Electrical Exhibition, Paris 1881 26
Gunpowder – see Black powder
Guns lost and destroyed in action 55, 116, 155

Hand-cranking 7, 9, 12, 16-18, 20, 28
Hang-fires 18, 21
Hazleton, Lt Cdr George 90
Hewitt, Edward 26
Holliday, Pte Sidney 151
Home Guard 96, 107
Hotchkiss guns 28, 36, 89, 94
Hug-Chang, Li 37

Ignition 13
India 8
Indian Army 65
Institution of Mechanical Engineers, London 27
International Inventions Exhibition, Crystal Palace 1885 31

Jackman, Capt James 156

Kaiser Wilhelm I 40
Kenya uprising 1954 97
King George V 150
King Henry VIII 13
Kitchener, Lord 53
Korean War 97, 111-112, 161

Labortatoire Central des Poudres et Salpêtres, Paris 21
Laddis, Sgt Dave 161
Le Brun, Reginald 152-153
Lees, Cpl Barry 160
Lefaucheux, Casimir 14
Lewis guns 32, 55, 59, 79, 87, 89, 92, 94-95, 104, 114, 153, 155
Lillie, Sir James 15-16
  Battery Gun 15
Lincoln, President 9, 16-17
Lithgow Vickers, Australia
  guns 56, 109-113
  Pratt & Whitney tooling 109
Lloyd George, David 52-53
Lobengula, King of Matabele 148
Loewe, Ludwig 40

Loewe, Sigmund 37, 40, 43-44
Longfield, Sgt 60

Maintenance of guns 119-145
  cleaning 65-67, 72, 121, 139
  clearing plugs 66-67
  in desert conditions 129
  in extreme cold 65
  lubrication and oiling 67, 108, 112, 123, 129, 141, 158
  sleeping with the gun 65
  timing check 125
Maxim guns
  air-cooled 36
  auto-loader 25
  first use in combat 148
  Forerunner blowback gun 30
  M1893 Service Maxim .303 88, 27, 30, 38, 40, 42, 44, 51, 65, 74, 76, 80
    trajectory table 76
  M1894 40
  M1895 Extra Light Rifle Calibre 36, 41, 43
  M1904 104, 106
  M1915 106
  Perfect Models 9, 36-37
  prototype machine gun 27, 31-32
  Quick-Firing One-Pounder Autocannon (Pom-Pom) 36, 39, 53, 102, 148
  test-firing 30
  Transitional Model 33, 36
  World Standard gun 33, 39, 41, 43
Maxim Gun Co. 42
Maxim, Hiram Stevens 9 *et seq.*
  apprenticeship 24
  autobiography 148
  becomes British subject 43
  birth 24
  death 163-164
  designs and patents 25-26, 30, 33, 43
  fascination with electrical energy 25
  genius 9, 26
  honours and wealth 163
  inventions 24-25
    current regulator 25-26
    electric lamp 25-26
  invited to Switzerland 37
  knighted 43
  Légion d'Honneur award 26
  London lecture 1855 19
  portraits 24, 32, 163
  sent to Europe 26
  severs ties with Maxim company 43
Maxim, Levi 24
Maxim-Nordenfelt guns 32, 38, 41, 51, 111
Maxim-Weston Co. 28
McWinney, Rindge & Co. 20
Ministry of Defence 97
Ministry of Munitions 52, 59
Montiny, Joseph 9
Moreau, Janne 165
Mortars 97, 160-161
  Grenatenwerfer 81mm 96
Motorcycles and sidecars 82, 93
  Clynos 82, 113-114
  Scott 87
Mounts – see also Tripods 38
  Australian 108, 110
  dovetail 100
  Mk I 50
  monopods 71
  Mounting, Trench, MG Mk I 60
  naval quad 103
  pintle 17, 38, 103, 114, 141
  Sangster bipod 59, 62, 71, 86, 110, 152-155

Scarff ring 92
Soklov 117
vehicle 102
wheeled armoured 39
Muskets 12-13
snaphaunce 12-13
Muzzle flash 9, 25, 61, 66, 159
Muzzle-loading 13-14
Myers & Son, Birmingham 58

Natal Carbineers 43
National Firearms Collection 12
Nordenfelt guns 18-19, 28, 37-38, 42, 75
Erith Works 39
four-barrel 1in calibre 17, 27
Nordenfelt, Thorsten 18, 38

Owen sub-machine gun 161

Palmcrantz, Helge 18, 75
Pauly, Jean Samuel 14
Patents 16, 25-26, 30, 33
Percussion caps 14, 21, 24
Periscopes 59-60, 66
Youlton Hyposcope 59-60, 66
Philip V of Spain 13
Pillbox emplacements 60
muzzle fumes sucked in 61
Pistols 12
flintlock revolving 12
multiple-barrelled 12
Portsmouth Naval Dockyard 18
Priming compound 14
Prince Esterhazy, HRH 37
Production, Maxim/Vickers
guns 59, 80, 104
Argentina 40
Bath shadow factory 70
Birmingham 40
cessation of manufacture 155
Colt Co., USA 105-106
Crayford works 38-40, 43, 49, 52-54, 70, 101, 110, 155
Dartford 40
employees 40, 52-55
women 53-55
laid off 155
Erith works 38-40, 43, 48, 52-55, 70, 110, 155
export sales 103-104, 107, 116
Hatton Garden, London workshops 31
last gun supplied to British Army 96
licensed production 40, 109, 112, 116
orders and contracts 37, 39-40, 50, 52-55, 70, 117
manufacture resumed 1930s 70
Russia 50, 117
prices 52, 55
production figures 41, 49-55, 70, 102, 105-107, 110, 117, 164
Russian M1910 116
US M1915 104-107
Propellant 12, 14, 21, 29
gas 20, 25
nitrocellulose 21, 32
Poudre B (Poudre Blanc) 21
Puckle, James 7, 13
Puckle gun 12
Defence gun 12-13
Puckle guns 12-13
Punch magazine 50

Queen Victoria 43

Rate of fire 16-17, 20, 31, 37, 51, 86, 89-90, 93, 102, 121, 135
Receiver 45, 56, 139
Recoil 16, 25, 30, 39, 59, 86, 90, 123
harnessing 29
Requa, Dr Josephus 16
Revolvers 7, 12, 15
Annely 14
Webley .455 91, 129
Rickenbacher, Eddy 92
Rifles 21, 41, 109
bolt-action magazine 21
Enfield 59
1853 Pattern 41
Fusil Modèle (Lebel) 21
Lee-Enfield 51, 55, 109
Martini-Henry 148
semi-automatic 29
Springfield .58 24
Thompsons M1928-A1 109
Winchester Model 1873 29
Ripley, Ezra 16
Ripley, Gen J.W. 16
Robert, Louis-Nicolas 8
Royal Air Force (RAF) 89
Royal Armouries 12
Royal Artillery 58, 67, 77, 87
Royal Artillery Museum, Woolwich 15
Royal Flying Corps (RFC) 89, 91
Royal Laboratories, Woolwich 75
Royal Naval Air Service (RNAS) 94
Royal Navy 18, 41, 75, 102
Royal Small Arms Factory (RSAF), Enfield 41, 70, 101
refurbishing old guns 115
.45 Gatling Calibre gun No. 1 41
Russian Revolution 1917 107, 117
Russo-Japanese War 148-149

Sangster, Charles 59, 62
Sangster, Thomas B. 62
SAS 158
Second World War 64, 96, 101, 155
Australian gun and spares production 110-111
Belgium 160
Borneo 72
British Expeditionary Force (BEF) 155-156
Burma 158
Chindit units 158
Dieppe Raid 116
Dunkirk retreat 156
El Alamein 73
France 155
Allied invasion, 1944 159-160
German Army sweep through Low Countries and France 155
Greece 112
Italy 112, 157-158, 160
Maas-Scheide Canal 159
Netherlands 160
New Guinea 112, 158
North Africa 112, 157
Operation Varsity 160
Overloon, Holland 71
Pacific 111
Sicily 157
South Pacific 112
Tobruk 156-157
Western Desert 111, 156
Seton-Hutchinson, Lt Col Graham 83, 152
Shaw, Joshua 14
Sights and rangefinding 37, 76-77, 84
angle-of-sight instrument 77
anti-aircraft 67, 81

bar foresight 76-77
clinometers 59, 67, 76-77, 84
compasses 84
dial sights 59, 76-77, 159
optical 58-59, 92
manufacturers 58, 67
Mk II 77
Peddie-Calochiopulo rearsight 39, 51
prism 60
ramp-type 48
rangefinders 67, 77, 80, 84, 97
rear 91, 97, 100, 105-106, 150
ring and bead 89
slide rules 77
tangent sights 70
Silverman, Louis 33
Small Arms School, Hythe 51, 67, 80
Smith & Wesson No. 1 revolver 15
Smith, Horace 8, 15
Snipers 59, 67, 75, 154-155
Spandau Arsenal 40
Spares 59, 86, 109-111, 113, 133
from dismantled guns 116
leather wallet 66
small parts tin 58
Specifications 169
Vickers High-Velocity automatic gun 12.7mm 169
Vickers Mk I .303mm 169
Vickers 0.5in Mk III 169
weight 49, 100, 102, 117
Spencer, Henry and Sharps 14
Springfield Armory 104

Tools 58, 66-67
belt expander 61, 73
Plug, Clearing Gun 143
Training 50, 80-81, 83, 87, 95-96, 107, 109, 111-112, 141, 155
aircraft use 89
Belton Park, Grantham 83
Blacon, Cheshire 95, 161
circle of guns 169
Handbook 96
Netheravon, Wiltshire 95-96, 161
Saighton, Cheshire 95, 161
Wisques machine gun training school 81
Transport of guns, tripods and ammunition
dropping by parachute 159
horses and mules 86, 108, 158-159
motor road vehicles 161
wooden carrying case 67
Trials, testing and demonstrations 37, 40, 45, 49, 51, 101-105
Tripods 13, 33, 49-51, 60, 67, 86-87, 110, 138-144, 149
armoured 32
carrying 133, 149, 155, 158
Class C 104
Colt-Vickers 105-106, 112
common problems 127
J-pattern 54
Mk I 8, 41
Mk II 50
Mk IV 51, 56, 59, 64, 70-71, 110, 112, 127, 141, 152
direction dial 84
production figures 59
VSM Mk B 44
Tula Arsenal, Russia 117

Unge, Eric 18
United States Electrical Lighting Co. 26
US Army 9, 16, 18, 25, 107
US Army Air Force 107

US Army Ordnance, Washington Arsenal 18
US Board of Ordnance 16, 104-105
US Model 1915 Vickers gun 104-107
US Navy 41
US Patent Office 16

Vickers 38, 40, 42, 70, 101
acquired Supermarine plant 70
merged with Armstrong Whitworth 70
tank production 101
Vickers, Albert 36-37, 43
Vickers-Berthier gun 104
Vickers, Tom 43
Vickers Gas-Operated aircraft gun 32
Vickers-Maxim and Vickers guns 40, 44, 48, 51
Class C 104
Class D 102
Class K aircraft gun 104
improvements and modifications 43-44, 111
last use by British Army 9, 161
Mk I 49, 52, 80, 82, 88-89, 99, 101, 110,
.5in AFV gun 100-101, 103
Mk II 70, 88, 93, 102
Mk III 102
Mk IV 102
Mk IVA 100
Mk IVB 100
Mk V 102
Mk VII 101
Mk XXI 110-111
model designations 45
M1904 New Pattern 105
M1906 New Light gun 44, 49
M1908 Light gun 39
Model (Class) C 40, 45, 50
Model E 50
Model F 50
QF 6-pounder 101
Russian 116-117, 148-149
Type B 45
US M1915 104-107
Vickers Sons & Maxim (VSM) 43
Vieille, Paul Marie 21

Waring & Gillow 58
War Department 59
War Office 39, 43, 49-52, 54
Water-jacket cooling 18, 31, 40, 43, 45, 51, 58, 64-65, 79, 91, 94-95, 100, 104, 106, 110, 117
anti-freeze 97
condenser bag 129
draining 121, 129
fluted 44-45, 65, 104
freezing 92, 117
funnels 64
hoses and connectors 64, 101, 105, 129, 139
lack of coolant 129
leather and canvas covers 42, 71
repair kits 64-65
shrapnel punctures and repairs 65, 121
slipstream cooling 89
smooth steel 48, 65, 70, 101, 110-111
topping up 137
tractor cap filler 117
Wesson, Daniel B. 8, 15
W.G. Armstrong & Co. 36
Williamson, Pte Thomas 154-155
Winborg, Johan Théodor 18
Wood, Francis Derwent 165
Woolwich Arsenal 15, 37